What Every Manager
Should Know About Quality

What Every Manager Should Know About Quality

Thomas Pyzdek
Quality America, Inc.
Tucson, Arizona

Marcel Dekker, Inc.   New York · Basel · Hong Kong
ASQC Quality Press   Milwaukee, Wisconsin

Library of Congress Cataloging-in-Publication Data

Pyzdek, Thomas.
    What every manager should know about quality / Thomas Pyzdek.
      p. cm.
    Includes bibliographical references (p.) and index.
    ISBN 0-8247-8401-4
    1. Production management--Quality control.   I. Title.
TS156.P993   1990
658.5'62--dc20   90-49144
                                                CIP

This book is printed on acid-free paper.

Marcel Dekker, Inc.
270 Madison Avenue, New York, New York 10016

Current printing (last digit):
10 9 8 7 6 5 4 3 2 1

PRINTED IN THE UNITED STATES OF AMERICA

Preface

After publishing *What Every Engineer Should Know About Quality Control* I was inundated with requests for a book directed toward management. After all, the readers argued, most of the quality problems can only be solved by management. They pointed out that by directing quality books at engineers, operators, and quality control people I was "preaching to the choir." Management must be included if real progress is to be possible.

This book is adapted from *What Every Engineer Should Know About Quality Control.* I stated in the preface to that book that I believe every engineer should be an *expert* on quality. I believe this even more strongly when it comes to managers. Quality improvement is much more a management responsibility than an engineering one. Obviously, reading a single book will not make you an expert. However, this book will give you a sample of the most important areas—enough to be conversant with the subject and, I hope, enough to stimulate you to learn more. In short, I have opted for breadth over depth. The treatment is not technical. Quality is best achieved by focusing on the fundamentals.

Another objective is to give you enough information to provide the essence of what quality control involves. The book provides a brief overview of the disciplines affecting quality control, with enough detail to give you an accurate feel for how each discipline is actually applied to quality control work.

Quality has always been an issue of importance. Many experts believe that America's failure to maintain high levels of quality is largely responsible for her declining position in world markets. The *Harvard Business Review* concluded in 1983 that if America did not dramatically improve quality " . . . our whole economy comes into jeopardy." A report by the MIT Commission on Industrial Productivity released in June 1989 concluded that "Simply put, many U.S. firms have been outperformed in the design and manufacture of reliable, high-quality products." Clearly, America *must* improve quality dramatically and quickly. Management plays the key role in this effort. It follows that management must develop a much better understanding of quality.

Recent developments have made quality much more a science than an art form. The body of knowledge of quality improvement is now well defined, and it is becoming increasingly sophisticated. Today's issues of consumerism, product and professional liability, and government regulation make formal consideration of quality an absolute necessity.

Quality is more than simply "goodness." Quality control can be thought of, broadly, as the science of discovering, controlling, and reducing variation. It is a body of specialized knowledge that encompasses disciplines as diverse as mathematics and management, psychology and engineering, and the law, and human relations. A central theme of quality control is that carefully conceived and well-engineered designs should be consistently reproduced. The importance of this to every manager is obvious.

This book represents the vital elements from the vast body of knowledge about the most important quality management methods and tools. By understanding the science of quality managers will better appreciate their place in the overall picture. In the future quality will become ever more important, and managers will be asked for their contribution to quality. The ability to make a contribution will depend on how well the basics are understood.

This book is *not* a complete treatise on the subject of quality—far from it. Quality is a vast field that includes contributions from many different disciplines. My intention is to provide an overview of the most important elements, occasionally providing enough detail to give you the "feel" of an especially important technique.

I believe that most books on quality are too sterile and academic. Therefore, I have not spared you my opinions. My work in the field has been educational, interesting, and exciting, and I wouldn't want to deprive you of the insights I've acquired along the way! I hope that you find my editorial digressions enlightening, entertaining, and educational.

Thomas Pyzdek

Contents

What Every Manager
Should Know About Quality

1
Basic Concepts of Quality Control

WHAT IS QUALITY?

Webster's New 20th Century Dictionary defines *quality* as "the degree of excellence that a thing possesses." Unfortunately, while this is the common usage of the term, it is useless as a working definition. If you were given a set of fine china, chances are you'd agree that the dinnerware was "quality merchandise." But what if you were asked to judge the quality of a set of paper plates? Although it is obvious that the paper plates don't compare to fine china, it's equally clear that the users of these products have certain things in mind when they make their purchase. The fine china would be as out of place on a picnic as the paper plates would be at a formal dinner. Thus, if we want a definition of quality that works with paper plates as well as with fine china, we must look beyond the dictionary.

The quality profession has struggled with the definition of quality for quite some time. Some of the most respected people in the quality profession have advanced their own definitions. Dr. W. Edwards Deming defines two different types of quality. Quality of *conformance* is the extent to which a firm and its suppliers surpass the design specifications required to meet the customer's needs. Quality of *performance* is the measure, determined through research and sales/service call analysis, of how well products perform in the marketplace. This leads to redesign, new specifications, and never-ending improvement.

Other quality experts have other definitions. Philip Crosby says quality is conformance to requirements, period. Joseph Juran defines quality as fitness for use. To Armand Feigenbaum, quality is "the total composite product and service characteristics of marketing, engineering, manufacture, and maintenance through which the product and service in use will meet the expectations of the customer." Unlike the dictionary definition, these definitions will provide a basis for developing a quality control system. It remains, however, for us to determine what is meant by "control."

WHAT IS CONTROL?

Once again, let's start with Webster. One definition of control is "a standard of comparison for verifying or checking the findings of an experiment." This time Webster comes quite close to providing a definition we can work with, as can be seen by comparing this to Juran's definition of control: "The process of measuring quality performance, comparing it to requirements, and acting on the difference."

Feigenbaum defines control as "a process for delegating responsibility and authority for a management activity while retaining the means of assuring satisfactory results." Feigenbaum's definition is sufficiently generic to apply to any activity, not just control of quality. This can be seen even more clearly if we look at Feigenbaum's four steps of control:

1. Setting standards
2. Appraising conformance
3. Acting when necessary
4. Planning for improvements

THE QUALITY CONTROL SYSTEM

At one time quality control was viewed, quite narrowly, as product inspection. "Quality control" consisted of assigning the last person on the assembly line the task of ensuring that the product worked. In modern practice, quality control begins with the design process and continues through manufacturing and use of the product. The sum of all of these efforts is called *total quality control (TQC)*. Quality control can also be viewed as all activities directed toward discovering and controlling variation in performance. The principles of quality control can be applied equally well to either products or services. We always seek, as an ultimate goal, products and services of consistent excellence. Futhermore, each unit or individual within a company can be viewed as providing a product or service for some other individal or unit, and the output can be evaluated using the tools of quality control.

Total quality control encompasses all eight stages of the product life cycle (Feigenbaum):

1. Marketing
2. Engineering
3. Purchasing
4. Manufacturing engineering
5. Production
6. Inspection and test
7. Shipping
8. Installation, maintenance, and service

Obviously, such a broad scope involves nearly everyone in the organization. The idea that the entire organization must be involved in the quality control effort is not new; quality experts advanced the idea as early as 1951, and the quality movement has been evolving toward TQC almost since the dawn of the industrial age. However, TQC stalled in America during the 1950s, when the quality effort was delegated to specialists in separate quality departments. Japan, on the other hand, openly embraced the TQC concept and proved that it works. This has resulted in a revival of the approach in the United States.

Quality planning is at the heart of TQC. Quality planning is an activity directed toward *preventing* quality problems. The concerns addressed by quality planning include

Establishing quality guidelines
Building quality into the design
Procurement quality
In-process and finished product quality
Inspection and test planning
Control of nonconforming material

Handling and following up on customer complaints
Education and training for quality

Establishing Quality Guidelines

Quality guidelines are established by determining customer requirements. Until the customer's wants and needs are clearly understood, it is impossible to develop a meaningful quality plan. Historically, customer needs have been determined by marketing and sales groups, often working independently of other activities within the company. Recently, however, some companies have involved production, engineering, and other groups in the effort by forming interdisciplinary teams. The joint effort has resulted in better understanding of the customer's true wants and needs and, ultimately, in better quality designs and finished goods. Once the customer's needs are known, management must determine whether company policies, procedures, and objectives are consistent with those needs. If not, they must be revised.

Building Quality into the Design

Design quality is addressed by comparing proposed design concepts with the needs of the customer, including reliability and maintainability considerations. At the same time, the design is reviewed for producibility and inspectability. It is possible to design a product that meets all of the customer's requirements but can't be made with existing technology. Sometimes the product is flawed by the use of exotic materials that are extremely costly or difficult to obtain. It is also possible to design a product that can be built, but can't be tested or inspected. If this occurs, the customer becomes the inspector, which is usually unacceptable. The design review process typically involves several departments, including marketing, manufacturing, purchasing, and quality control. Design review should be a formal activity, with mandatory sign-offs and documentation.

Procurement Quality Function

Since most buyer-seller arrangements involve contracts, the procurement quality activity tends to be quite formal. However, the importance of the less legalistic aspects of the relationship should not be overlooked. A loyal supplier, committed to your success, is something you can't get with a contract. The procurement quality function is so important that Chapter 2 of this book is devoted entirely to the subject.

In-Process and Finished Product Quality Planning

This activity involves establishing specifications for all important quality characteristics and developing formal product standards. Work instructions and detailed operating procedures are also part of this effort.

Inspection and Test Planning

Inspection and test planning is carefully integrated with the design and production activities. The work involves determination of inspection stations, classification of characteristics according to their criticality, design and procurement of inspection and test equipment, and development of inspection instructions and test procedures. These activities are typically performed by quality control or test specialists.

Control of Nonconforming Material

Nonconforming materials must be carefully segregated and identified to prevent their inadvertent mixing with acceptable materials. The material control procedures are often discussed in the purchase order or contract. Typically, a material review board composed of manufacturing, engineering, quality control, and, perhaps, a customer representative periodically reviews nonconforming materials and makes the final decision on their disposition (e.g., sort, scrap, rework). It is important to recognize that the nonconforming material is a *symptom* of some problem, and corrective action necessary to prevent future nonconforming material must be an essential element of the nonconforming material control system.

Customer Complaints

Data from customer complaints must be taken very seriously. In fact, customer complaints should be actively solicited. A major auto company discoved that a dissatisfied customer tends to complain to 14 people about his or her experience. A highly satisfied customer, on the other hand, tells only 7 others. The importance from a marketing point of view is obvious. Customer complaints are also a valuable source of information for future product improvements. Customer complaints involve two separate elements: correcting the customer's immediate problem and utilizing the data for long-term improvements.

Education and Training for Quality

W. Edwards Deming once said that most people know everything they need to know about their jobs, except how to do them better. Training in the tools of quality control, especially statistical analysis, is essential to the quality improvement effort. Quality is also affected by the amount of on-the-job training people receive, as well as by special training in job-related skills. Training is an ongoing activity that should begin before an employee is hired and continue throughout the employee's term of service.

INSPECTION AND TESTING

Inspection and testing activities have a direct impact on the quality of the product delivered to the end user. While it may be true that "you can't inspect quality into the product," you can inspect bad quality *out* of the product. Inspection and test are so important to quality that at one time they *were* quality control.

Inspection and test activity is usually directed toward one of the following goals:

Determination of product acceptance relative to
customer requirements
Determination of product reliability
Qualification of a vendor, process, machine, etc.
Verification that some requirement has been met

Inspection and test involve cost. With most products, the cost of inspecting every characteristic of every unit is prohibitive. One way of dealing with this is to classify characteristics of the product and devote the bulk of attention to the characteristics most important in meeting the customer's requirements. Typically, three classifications are used: critical, major, and minor. A critical characteristic is one that has a direct impact on the safe use of the product. A major characteristic is one that is not critical but is likely to affect the usability or performance of the product. A minor characteristic is one unlikely to affect the performance of the product.

Another way to deal with the issue of economics is to sample rather than inspect 100% of the units produced. Sampling involves making decisions based on the results of inspections or tests on less than every unit. The benefit from sampling is obvious: lower costs. The opposite side of the coin is that you must tolerate a level of risk, known as sampling risk. There are two kinds of sampling risk: good products may be classified incorrectly as nonconforming (producer's risk), or nonconforming products may be classified incorrectly as conforming (consumer's risk). When dealing with critical characteristics there is no acceptable level of consumer risk, and therefore you should *never* sample critical characteristics except to verify a previous 100% inspection. An exception to this arises when the inspection or test is destructive, in which case you have no option but to sample. Once you decide to sample, you must face the question of how much to sample. This subject will be discussed in detail in subsequent chapters.

All inspection and test activity involves some degree of error. No measurement instrument is perfect, and human beings are fond of quoting the phrase " to err is human" for good reason. Interpretation of inspection and test results should be tempered by this knowledge. There are statistical methods of evaluating all sorts of inspection and test errors, but many of them are beyond the scope of this book. However, certain types of errors can easily be evaluated and controlled. For example, the accuracy of a micrometer can be checked by measuring a part of known size and adjusting the micrometer until it gives the correct result. The "part of known size" is called a standard, and the adjustment of the gage is called calibration.

Certain types of tests are classified as either destructive or nondestructive. While any test that does not do damage to the product or raw material is technically nondestructive, in practice the term *nondestructive test* is reserved for certain test procedures. The test methods most commonly classified as nondestructive are eddy current, dye penetrant, magnetic particle, ultrasonic, and X-ray tests. These tests are most often used to check properties of materials or products such as cracks and porosity.

QUALITY AND THE LAW

The subject of quality and the law is also known as product liability. We will provide only a brief overview of the subject. It is so important to the engineering profession that Volume 2 of the *What Every Engineer*

Should Know series (Marcel Dekker, Inc.) is devoted entirely to the subject. Table 1.1, reproduced from that volume, defines the legal terminology important in quality control.

Three legal theories are involved in product liability: breach of warranty, strict liability in tort, and negligence. Two branches of law deal with these areas, contract law and tort law. We will review the ways in which these legal concepts affect quality control.

A *contract* is a binding agreement for whose breach the law provides a remedy. In quality control work, as in product liability in general, the contract is related to the sale of a product. Contract laws for such cases are usually governed by the Uniform Commerial Code (UCC). Key concepts of contract law related to product liability are those of *breach of warranty* and *privity of contract*.

Breach of warranty can occur from either an express warranty or an implied warranty. An express warranty is part of the basis for a sale. In other words, the buyer agreed to the purchase on the reasonable assumption that the product would perform in the manner described by the seller. The seller's statement need not be written for the warranty to be an express warranty; a mere statement of fact is sufficient. An implied warranty is a warranty not stated by the seller but implied by law. Certain warranties result from the simple fact that a sale has been made. One of the most important of the attributes guaranteed by an implied warranty is that of fitness for normal use. The warranty is that the product is reasonably safe.

Privity of contract means that a direct relationship exists between two parties, typically buyer and seller. At one time, manufacturers were held not liable for products they purchased from vendors or sold to a consumer through a chain of wholesalers, dealers, etc. Manufactures were treated as third part assignees and said to be not in privity with the end user. In contract law, privity denotes parties in mutual legal relationship to one another by virtue of being promisees and promisors. This concept began to deteriorate in 1905, when courts began to permit lawsuits against sellers of unwholesome food, whether or not they were negligent, and against original manufacturers, whether or not they were in privity with the consumers. The first recognition of strict liability for an express warranty without regard to privity was enunciated by a Washington court in 1932 in a case involving a Ford Motor Company express warranty that their windshields were "shatterproof." When a windshield shattered and injured a consumer, the court

Table 1.1 Fundamental Legal Terminology

Assumption of risk: The legal theory that a person who is aware of a danger and its extent and knowingly exposes himself to it assumes all risks and cannot recover damages, even though he is injured through no fault of his own.

Contributory negligence: Negligence of the plaintiff that contributes to his injury and at common law ordinarily bars him from recovery from the defendant although the defendant may have been more negligent than the plaintiff.

Deposition: The testimony of a witness taken out of court before a person authorized to administer oaths.

Discovery: Procedures for ascertaining facts prior to the time of trial in order to eliminate the element of surprise in litigation.

Duty of care: The legal duty of every person to exercise due care for the safety of others and to avoid injury to others whenever possible.

Express warranty: A statement by a manufacturer or seller, either in writing or orally, that his product is suitable for a specific use and will perform in a specific way.

Foreseeability: The legal theory that a person may be held liable for actions that result in injury or damage only when he was able to foresee dangers and risks that could reasonably be anticipated.

Great care: The high degree of care that a very prudent and cautious person would undertake for the safety of others. Airlines, railroads, and buses typically must exercise great care.

Implied warranty: An automatic warranty, implied by law, that a manufacturer's or dealer's product is suitable for either ordinary or specific purposes and is reasonably safe for use.

Liability: An obligation to rectify or recompense for any injury or damage for which the liable person has been held responsible or for failure of a product to meet a warranty.

Negligence: Failure to exercise a reasonable amount of care or to carry out a legal duty which results in injury or property damage to another.

Obvious peril: The legal theory that a manufacturer is not required to warn prospective users of products whose use involves an obvious peril, especially those that are well known to the general public and that generally cannot be designed out of the product.

Prima facie: Such evidence as by itself would establish the claim or defense of the party if the evidence were believed.

Table 1.1 *(Continued)*

Privity: A direct contractual relationship between a seller and a buyer. If A manufactures a product that is sold to dealer B, who sells it to consumer C, privity exists between A and B and between B and C, but not between A and C.

Proximate cause: The act that is the natural and reasonably foreseeable cause of the harm or event that occurs and injures the plaintiff.

Reasonable care: The degree of care exercised by a prudent person in observance of his legal duties toward others.

Res ipsa loquitur: The permissible inference that the defendant was negligent in that "the thing speaks for itself" when the circumstances are such that ordinarily the plaintiff could not have been injured had the defendant not been at fault.

Standard of reasonable prudence: The legal theory that a person who owes a legal duty must exercise the same care that a reasonably prudent person would observe under similar circumstances.

Strict liability in tort: The legal theory that a manufacturer of a product is liable for injuries due to product defects, without the necessity of showing negligence of the manufacturer.

Subrogation: The right of a party secondarily liable to stand in the place of the creditor after he has made payment to the creditor and to enforce the creditor's right against the party primarily liable in order to obtain indemnity from him.

Tort: A wrongful act or failure to exercise due care, from which a civil legal action may result.

---

allowed the suit against Ford, ruling that even without privity the manufacturer was responsible for the misrepresentation, even if the misrepresentation was made innocently.

Under the rule of *strict liablity*, an innocent consumer who knows nothing about disclaimers and the requirement of giving notice to a manufacturer with whom he did not deal cannot be prevented from suing. The rule avoids the technical limits of privity, which can create a chain of lawsuits back to the party that originally put the defective product into the stream of commerce. The seller (whether a salesperson or manufacturer) is liable even though he has been careful in handling the product and even if the consumer did not deal directly with him.

This modern rule was first applied in the case of *Greenman v. Yuba Power Products, Inc.* in California in 1963. A party, Mr.Greenman, was injured when a workpiece flew from a combination power tool purchased for him by his wife two years prior to the injury. He sued the manufacturer and produced witnesses to prove that the machine was designed with inadequate setscrews.

The manufacturer, which had advertised the power tool as having "rugged construction" and "positive locks that hold through rough or precision work," claimed that is should not have to pay money damages because the plaintiff had not given it notice of breach of warranty within reasonable time as required. Futhermore, a long line of California cases held that a plaintiff could not sue someone not in privity with him unless the defective product was food.

The court replied that this was not a warranty case but a *strict liability case.* The decision stated that any "manufacturer is strictly liable ... when an article he placed on the market, knowing that it is to be used without inspection for defects, proves to have a defect that causes injury to a human being."

The concept of strict liability was a turning point for both the consumer movement and quality control. The use of effective, modern quality control methods became a matter of paramount importance. The concept is also called *strict liability in tort,* which is virtually synonymous with the common usage of the term "product liability." A *tort* is a wrongful act or failure to exercise due care resulting in an injury, from which civil legal action may result. Tort law seeks to provide compensation to people who suffer loss because of the dangerous or unreasonable actions of others.

A related concept is that of *negligence.* Negligence occurs when one person fails to fulfill a duty owed to another or fails to act with due care. Two elements are necessary to establish negligence: a standard of care recognized by law, and a breach of the duty or requisite care. Also, the breach of duty must be the proximate cause of the harm or injury. The accepted standard of care is that of the "reasonable person." The court must measure the action of the parties involved relative to the actions expected from an imaginary reasonable person.To muddy the waters futher, the court must weigh the risk or danger of the situation against the concept of "reasonable risk." Clearly, these concepts are far from cut and dried.

The case cited above and many other developments since it have resulted in a feature that is unique to product liability law, namely: *the conduct of the manufacturer is irrelevant.* The plaintiff in a product liability suit need not prove that the manufacturer failed to exercise due care; he need show only that the product was the proximate cause of harm and that it was either defective or unreasonably dangerous. This is what is meant by strict liability. In a sense, it is the product that is on trial and not the manufacturer. The concept is defined formally as follows:

1. One who sells any product in a defective condition unreasonably dangerous to the user or consumer or to his property is subject to liability for physical harm thereby caused to the ultimate user or consumer, or to his property, if
    a. The seller is engaged in the business of selling such a product, and
    b. It is expected to and does reach the user or consumer without substantial change in the condition is which it is sold.
2. The rule in subsection (1) applies although
    a. The seller has exercised all possible care in the preparation and sale of the product, and
    b. The seller of consumer has not bought the product from or entered into any contractual relation with the seller.

As discussed in Volume 2 of this series, there are several areas in which engineering and management are vulnerable:

1. Design
2. Manufacturing and materials
3. Packaging, installation, and application
4. Warnings and labels

Designs that create hidden dangers to the user, designs that fail to comply with accepted standards, designs that exclude necessary safety features or devices, or designs that do not properly allow for possible unsafe *misuse or abuse* that is reasonably foreseeable to the designer are all suspect. Quality control includes design review as one of its major elements, and all designs should be carefully evaluated for these shortcomings. As always, the concept of reasonableness applies in all its ambiguity.

The application of quality control principles to manufacturing, materials, packaging, and shipping is probably the best protection possible against future litigation. Defect prevention is the primary objective of quality control, and the defect that isn't made will never result in loss or injury. Bear in mind, however, that a defect in quality control is usually defined as a nonconformance to requirements. There is no such definition in the law. Legal definitions of a defect are based on the concept of reasonableness and the need to consider the use and likely abuses of the product.

Quality control normally does not concern itself with warnings and labels. However, I have seen situations where required labels were missing or misplaced, labels didn't adhere to the surfaces to which they were affixed, and labels could not easily be seen by the end user. Such situations *are* in the domain of quality control.

SUMMARY

This chapter has introduced a working definition of quality and explained its importance. The concept of control was explained as it is related to quality control. Total quality control was introduced as an operating philosophy and guide for management. The importance of quality planning and product liability issues were discussed.

Several important topics were not covered in this chapter. The reader may wish to explore the different ways in which the basic concepts manifest themselves in different industries. Also of interest are the different applications of the basic principles in commercial firms and defense contractors. The relationship of the quality basics to important related areas such as safety, productivity, and employee relations is also worth further study.

2

Vendor Quality Assurance

The major part of the cost of most manufactured products is in purchased materials. In some cases the percentage is as high as 80%, and it is seldom less than 50%. The importance of consistently high levels of quality in purchased materials is clear. This chapter examines important aspects of vendor quality control systems.

ETHICS AND HUMAN RELATIONS

It is important to remember that dealings between companies are really dealing between *people*. People work better together if certain ground rules are understood and followed. Above all, the behavior of both buyer and seller should reflect honesty and integrity. This is especially important in quality control, where many decisions are "judgment calls." There are certain guidelines that foster a relationship based on mutual trust:

1. Don't be too legalistic. While it is true that nearly all buyer-seller arrangements involve a contract, it is also true that unforeseen con-

ditions sometimes require that special actions be taken. If buyer and seller treat each other with respect, these situations will present no real problem.

2. Maintain open channels of communication. This involves both formal and informal channels. Formal communication includes such things as joint review of contracts and purchase order requirments by both seller and buyer teams, on-site seller and buyer visits and surveys, corrective action request and follow-up procedures, and record-keeping requirements. Informal communications involve direct contact between individuals in each company on an ongoing and routine basis. Informal communication to clarify important details, ask questions, gather background to aid in decision making, and so on, will prevent many problems.

3. The buyer should furnish the seller with detailed product descriptions. This includes drawings, workmanship standards, special processing instructions, and any other information the seller needs to provide a product of acceptable quality. The buyer should ascertain that the seller understands the requirements.

4. The buyer should objectively evaluate the seller's quality performance. This evaluation should be done in an open manner, with the full knowledge and consent of the seller. The buyer should also keep the seller informed of his *relative standing* with respect to other suppliers of the same product. However, this should be done in a manner that does not compromise the position of any other seller.

5. The buyer should be prepared to offer technical assistance to the seller, and vice versa. Such assistance may consist of on-site visits by buyer or seller teams, telephone assistance, or transfer of documents. Of course, both parties are obligated to protect the trade secrets and proprietary information they obtain from one another.

6. The seller should inform the buyer of any known departure from historical or required levels of quality.

7. The buyer should inform the seller of any change in requirements in a timely fashion.

8. The seller should be rewarded for exceptional performance. Such rewards can range from plaques to increased levels of business.

The basic principles of ethical behavior have been very nicely summarized in the Code of Ethics for members of the American Society for Quality Control, reproduced in Figure 2.1.

The American Society for Quality Control
Code of Ethics

*To uphold and advance the honor and dignity of the profession, and in keeping with high standards of ethical conduct I **acknowledge that I:***

Fundamental Principles

I. *Will be honest and impartial, and will serve with devotion my employer, my clients and the public.*

II. *Will strive to increase the competence and prestige of the profession.*

III. *Will use my knowledge and skill for the advancement of human welfare, and in promoting the safety and reliability of products for public use.*

IV. *Will earnestly endeavor to aid the work of the Society.*

Relations With The Public

1.1 *Will do whatever I can to promote the reliability and safety of all products that come within my jurisdiction.*

1.2 *Will endeavor to extend public knowledge of the work of the Society and its members that relates to the public welfare.*

1.3 *Will be dignified and modest in explaining my work and merit.*

1.4 *Will preface any public statements that I may issue by clearly indicating on whose behalf they are made.*

Relations with Employers and Clients

2.1 *Will act in professional matters as a faithful agent or trustee for each employer or client.*

2.2 *Will inform each client or employer of any business connections, interests or affiliations which might influence my judgment or impair the equitable character of my services.*

2.3 *Will indicate to my employer or client the adverse consequences to be expected if my professional judgment is overruled.*

2.4 *Will not disclose information concerning the business affairs or technical processes of any present or former employer or client without his consent.*

2.5 *Will not accept compensation from more than one party for the same service without the consent of all parties. If employed, I will engage in supplementary employment of consulting practice only with the consent of my employer.*

Relations With Peers

3.1 *Will take care that credit for the work of others is given to those to whom it is due.*

3.2 *Will endeavor to aid the professional development and advancement of those in my employ or under my supervision.*

3.3 *Will not compete unfairly with others; will extend my friendship and confidence to all associates and those with whom I have business relations.*

Figure 2.1 ASQC Code of Ethics.

SCOPE OF VENDOR QUALITY CONTROL

Most companies purchase several different types of materials. Some of the materials are just supplies, not destined for use in the product to be delivered to the customer. Traditionally, vendor quality control does not apply to these supplies. Of the items destined for the product, some are simple items that have loose tolerances and an abundant history of acceptable quality. The quality of these items will usually be controlled informally, if at all. The third category of purchased goods involves items that are vital to the quality of the end product, are complex, and have limited or no history. Purchase of these items may even involve purchase of the vendor's expertise (designs, application advice, etc). It is the quality of this category of items that will be the subject of subsequent discussions.

VENDOR QUALITY SYSTEMS

One early question is always, "who is responsible for vendor quality?" When it is asked in this way, the result is usually much pointless discussion. Instead of beginning with such a broad question, it is usually better to break down the tasks involved in assuring vendor quality and then assign responsibility for the individual tasks. The chart below is one such breakdown (Juran, 1980, p. 229).

Responsibility Matrix

Department			
Product design	Purchasing	Quality control	Activity
X	X	XX	Establish a vendor quality policy.
	XX		Use multiple vendors.
X	X	XX	Evaluate quality capability of potential vendors.
XX		X	Specify requirements for vendors.
X		XX	Conduct joint quality planning.
	X	XX	Conduct vendor surveillance.
X		XX	Evaluate delivered product.
X	X	XX	Conduct improvement programs.
	XX	X	Use vendor quality ratings in selecting vendors.

X = shared responsibility.
XX = primary responsibility.

It is important to recognize that, in the end, the responsibility for quality always remains with the supplier. The existence of vendor quality systems for "assuring vendor quality" in no way absolves the vendor of responsibility. Rather, these systems should be viewed as an aid to the vendor.

MULTIPLE VENDORS

The subject of whether or not to use multiple vendors is one that arouses strong feelings both pro and con. For decades, the conventional wisdon in American quality control was that multiple vendors would keep all suppliers "on their toes" through competition. Also, having multiple vendors was looked at as a hedge against unforeseen problems like fire, flood, or labor disputes. These beliefs were so strong that multiple vendors became the de facto standard for most major government agencies, including the Department of Defense.

In the 1980s the consensus on multiple sources of supply began to erode. Japan's enormous success with manufacturing in general and quality in particular inspired American businesspeople to study the Japanese way of doing things. Among the things Japanese businesses do differently is that they *discourage* multiple source purchases whenever possible. This is in keeping with the philosophy of W. Edwards Deming, a noted American quality expert who provided consulting and training to many Japanese firms. The advocates of single-source procurement argue that it encourages the supplier to take long-term actions on your behalf and makes suppliers more loyal and committed to your success. A statistical argument can also be made: minimum variability in product can be obtained if the sources of variation are minimized, and different suppliers are an obvious source of variation.

Since both sides have obvious strong points, the decision regarding single-source versus multiple-source procurement is one that must be made on a case-by-case basis. Your company will have to examine its unique set of circumstances in light of the arguments on both sides. In most cases a policy of using single sources except under unusual conditions works well.

EVALUATING VENDOR QUALITY CAPABILITY

When making an important purchase, most companies want some sort of advance assurance that things will work out well. When it comes to quality, the vendor quality survey is the "crystal ball" used to provide this assurance. The vendor quality survey usually involves a visit to the vendor by a team from the buyer prior to the award of a contract; for this reason it is sometimes called *preaward survey*. The team is usually composed of representatives from the buyer's design engineering, quality control, production, and purchasing departments. The quality control elements of the survey usually include, at a minimum,

Quality management
Design and configuration control
Incoming material control
Manufacturing and process control
Inspection and test procedures
Control of nonconforming material
Gage calibration and control
Quality information systems and records
Corrective action systems

The evaluation typically involves use of a checklist and some numerical rating scheme. A simplified example of a supplier evalutation checklist is shown in Figure 2.2.

The checklist in Figure 2.2 is very simple compared to those used in practice. Many checklists run on to 15 pages or more. My personal experience is that these checklists are very cumbersome and difficult to use. If you are not bound by some government or contract requirement, I recommend that you prepare a brief checklist similar to the one shown and supplement it with a narrative that documents your personal observations. Properly used, the checklist can help guide you without tying your hands.

Bear in mind that a checklist can never substitute for the knowledge of a skilled and experienced evaluator. Numerical scores should be supplemented by the observations and interpretaions of the evaluator. The input of vendor personnel should also be included. If there is disagreement between the evaluator and the vendor, the positions of both sides should be clearly described.

In spite of their tremendous popularity, physical vendor surveys are only one means of evaluating the potential performance of a supplier. Studies of their accuracy suggest that they should be taken with a large grain of salt. One such study by Brainard (1966) showed that 74 of 151 vendor surveys resulted in incorrect predictions (i.e, either a good supplier was predicted to be bad or vice versa)—a coin flip would have been as good a predictor! I recommend that physical vendor surveys be avoided whenever possible. One option is to use so-called desk surveys. With desk surveys you have the prospective suppliers themselves submit lists of test equipment, certifications, organization charts, manufacturing procedures, etc. Check these documents over for obvious problems.

Vendor Evaluation Checklist	
criteria	result
The quality system is developed	
The quality system is implemented	
Personnel are able to identify problems	
Personnel recommend and initiate solutions	
Effective quality plans exist	
Inspection stations are identified	
Management regularly reviews quality program status	
Contracts are reviewed for special quality requirements	
Processes are adequately documented	
Documentation is reviewed by quality	
Quality records accurate and up to date	
Effective corrective action system	
Non-conforming material properly controlled	
Quality costs properly reported	
Changes to requirements controlled	
Adequate gage calibration control	

Figure 2.2 Vendor Evaluation Checklist. Since this type of evaluation is conducted at the supplier's facility, it is known as a physical survey.

Regardless of the survey method used or the outcome of the survey, it is important to keep close tabs on the actual quality history. Have the vendor submit "correlation samples" with the first shipments. The samples should be numbered and each should be accompanied by a document showing the vendor's quality inspection results and test results. Verify that the vendor has correctly checked each important characteristic and that your results agree, or "correlate." Finally, keep a running history of quality performance (see Quality Records later in this chapter). The best predictor of future good performance seems to be a record of good performance in the past. If you are a subscriber to the Government, Industry Data Exchange Program (GIDEP) you have access to a wealth of data on many suppliers. (GIDEP subscribers must also contribute to the data bank.) Another data bank is the Coordinated Aerospace Supplier Evaluation (CASE). If relevant to your application, these compilations of the experience and audits of a large number of companies can be a real money-and time-saver.

VENDOR QUALITY PLANNING

Vendor quality planning involves efforts directed toward preventing quality problems, appraisal of product at the vendor's plant as well as at the buyer's place of business, corrective action, dispostition of nonconforming merchandise, and quality improvement. The process usually begins in earnest after a particular source has been selected Most pre-award evaluation is general in nature; after the vendor is selected it is time to get down to the detailed level.

A first step in the process is the transmission of the buyer's requirements to the vendor. Even if the preliminary appraisal of the vendor's capability indicated that the vendor could meet your requirements, it is important that the requirements be studied in detail again before actual work begins. Close contact is required between the buyer and the vendor to assure that the requirements are clearly understood. The vendor's input should be solicited; could a change in requirements help the vendor produce better quality parts?

Next it is necessary to work with the vendor to establish procedures for inspection, test, and acceptance of the product. How will the product be inspected? What workmanship standards are to be applied? What in-process testing and inspection is required? What level of sampling will be employed? These and similar questions must be answered at this stage. It is good practice to have the first few parts completely inspected by both the vendor and the buyer to assure that the vendor knows which features must be checked as well as how to check them. The buyer many want to be at the vendor's facility when production first begins.

Corrective action systems must also be developed. Many companies have their own forms, procedures, etc. for corrective action. If you want your system to be used in lieu of the vendor's corrective action system, the vendor must be notified. Bear in mind that the vendor many need addtional training to use your system. Also, the vendor may want some type of compensation for changing his established way of doing things. If at all possible, let the vendor use his own systems.

At an early stage, the vendor should be made aware of any program you have that would enable him to reduce inspection. At times it is possible to certify the vendor so that *no* inspection is required by the buyer and shipments can go directly to stores, bypassing receiving inspection completely.

If special record keeping will be required, this must be spelled out. Many companies have special requirements imposed on them by the nature of their business. For example, most major defense items have traceability and configuration control requirements. Government agencies such as the Food and Drug Administration (FDA) often have special requirements. Automotive companies have record-keeping requirements designed to facilitate possible future recalls. The vendor is often in the dark regarding your special record-keeping requirements; it is your job to keep him informed.

POSTAWARD SURVEILLANCE

The focus up to this point has been on developing a process that will minimize the probability of the vendor producing items that don't meet your requirements. This effort must continue after the vendor begins production. However, after production has begun the emphasis can shift from an evaluation of systems to an evaluation of *actual* program, process, and product performance.

Program evaluation is the study of a supplier's facilities, personnel, and quality systems. While this is the major thrust during the preaward phase of an evaluation, program evaluation doesn't end when the contract is awarded. Change is inevitable, and the buyer should be kept informed of changes in the vendor's program. Typically, this is accomplished by providing the buyer with a registered copy of the vendor's quality manual, which is updated routinely (see Chapter 4 for futher information on quality manuals). Periodic follow-up audits may also be required, especially if product quality indicates a failure of the quality program.

A second type of surveillance involves surveillance of the vendor's process. Process evaluations involve a study of methods used to produce an end result. Process performance can usually be best evaluated by statistical methods, and it is becoming common to require that statistical process control (SPC) be applied to critical process characteristics (see Chapter 7). Many large companies require that their suppliers perform statistical process control studies, called process capability studies, as part of the preaward evaluation.

The final evaluation, product evaluation, is also the most important. Product evaluation consists of evaluating conformance to requirements. This may involve inspection at the vendor's site, submission of objective evidence of conformance by the vendor, inspection at the buyer's receiving dock, or actual use of the product by the buyer or the end user. This *must* be the final proof of performance. If the end product falls short of requirements, it matters little that the vendor's program look good or that all of the in-process testing meets established requirements.

Bear in mind that the surveillance activity is a communications tool. To be effective it must be conducted in an ethical manner, with the full knowledge and cooperation of the vendor. The usual business communication techniques, such as advanced notification of visits, management presentations, exit briefings, and follow-up reports, should be utilized to assure complete understanding.

SPECIAL PROCESSES

We will define a special process as one that has an effect that can't readily be determined by inspection or testing subsequent to processing. The difficulty may be due to some physical constraint, such as the difficulty in verifying grain size in a heat-treated metal, or the problem may simply be economics, such as the cost of 100% X-ray of every weld in an assembly. In these cases special precautions are required to assure that processing is carried out in accordance with requirements.

The two most common approaches to control of special processes are certification and process audit. Certification can be applied to the skills of key personnel, such as welder certification, or to the processes themselves. With processes the certification is usually based on some demonstrated capability of the process to perform a specified task. For example, a lathe may machine a special test part designed to simulate product characteristics otherwise difficult or impossible to measure. The vendor is usually responsible for certification. Process audit involves establishing a procedure for the special process, then reviewing actual process performance for compliance to the procedure. A number of books exist to help with the evalutation of special processes. In addition, there are inspection service companies from which you can hire experts to verify that special processes meet established guidelines. These companies employ retired quality control professionals as well as full-time personnel. In addition to reducing your costs, these companies can provide a level of expertise you may not otherwise have.

QUALITY RECORDS

Reference was made earlier in this chapter to the importance of tracking actual quality performance. In fact. this "track record" is the most important basis for evaluating a vendor and the best predictor of future quality performance. Such an important task should be given careful consideration. The purpose of vendor quality records is to provide a basis for action. The data may indicate a need to reverse an undesirable trend or may suggest a cost savings from reduced (or no) inspection. Also, the records can provide an objective means of evaluating the effectiveness of corrective action.

An early step is to determine what data will be required. Data considerations include analyzing the level of detail required to accomplish the objectives of the data collection described above (i.e., how much detail is needed to determine an appropriate course of action?). Some details normally included are part number, vendor identification, lot identification number, name of characteristic(s) being monitored, quantity shipped, quantity inspected, number defective in the sample, lot disposition (accept or reject), and action taken on rejected lots (sort, scrap, rework, return to vendor, use as is). These data are usually first recorded on a paper form and later entered into a computerized data base. The original paper documents are kept on file for some period of time, the length of time being dictated by contract, warranty, tax, and product liability considerations.

The data collection burden can be significant, especially in larger companies. Some companies have resorted to modern technology to decrease the work load. For example, common defects may have bar codes entered on menus for the inspector. The vendor will apply a bar code sticker with important information such as the purchase order number and lot number, which the computer can use to access a data base for additional details on the order. Thoughtful use of a bar code system can reduce the data input chore to a few keystrokes. Some bar code systems have been designed for use on personal computers, bringing the technology within the reach of nearly everyone.

Although the raw data will contain all of the information necessary to provide a basis for action, the information will be in a form that lacks meaning to the decision makers. To make the raw data meaningful, they must be analyzed and reported in a way that makes the information content obvious. Reports should be designed with the end reader in mind. The data reported should be carefully considered and only the necessary information should be provided. In most cases it will take several reports to adequately distribute the information the right people. Typically, the level of detail provided will decrease as the level of management increases. Thus, senior management will usually be most interest in summary trends, perhaps presented as graphs and charts rather than numbers. Operating personnel will receive reports that provide information on specific parts and characteristics. At the lowest level, audit personnel will receive reports that allow them to verify the accuracy of the data entered into the system.

The general subject of data collection and reporting is covered in great detail in Chapter 5.

VENDOR RATING SCHEMES

As the discussion so far makes clear, evaluating vendors involves comparing a large number of factors, some quantitative and some qualitative. Vendor rating schemes attempt to simplify this task by condensing the most important factors into a single number, the vendor rating, that can be used to evaluate the performance of a single vendor over time or to compare multiple sources of the same item.

Most vendor rating systems involve assigning weights to different important measures of performance, such as quality, cost, and delivery. The weights are selected to reflect the relative importance of each measure. Once the weights are determined, the performance measure is multiplied by the weight and the results totaled to get the rating. For example, we might decide to set up the following rating scheme:

Performance	Measure	Weight
Quality	% of lots accepted	5
Cost	Lowest cost/cost	300
Delivery	% on-time shipments	2

The calculations for three hypothetical suppliers are shown below.

Vendor	% lots accepted	Price	% on-time deliveries
A	90	$60	80
B	100	70	100
C	85	50	95

Vendor	Quality Rating	Price Rating	Delivery Rating	Total Rating
A	450	250	160	860
B	500	214	200	914
C	425	300	190	915

As you can see, the simple example above combines reject rates, delivery performance, and dollars into a single composite number. Using this value, we would conclude that vendors B and C are approximately the same. What vendor B lacks in the pricing category is made up for in quality and delivery. Vendor A has a much lower rating than B or C.

A valid question at this point would be, "is all of this discussion meaningful?" Obviously, one of the problems is the mixture of units. As one quality control text puts it, "apples + oranges = fruit salad." Other questions involve the selection of weights, disposition of rejected lots (scrap, rework, use as is, return to vendor, etc.), number of lots involved, cost of defects, types of defects, how late the deliveries were, whether early deliveries and late deliveries both count against the delivery score, and so forth. It is easier to answer such questions if some guidelines are provided. Here are some characteristics of good rating schemes:

> The scheme is clearly defined and understood by both the buyer and the seller.
> Only relevant information is included.
> The plan is easy to use and update.
> The scheme is applied only where it is needed.
> The ratings "make sense" when viewed in light of other known facts.

Finally, I must express some skepticism regarding rating schemes. I have personally never seen a rating scheme that didn't have many "holes" in it that required subjective interpretation which defeats the purpose of the rating. Also, there are very few cases where the ratings can be usefully applied: multiple sources of the same important item over a time period sufficient to provide a valid basis for comparison. Many ratings have values that are highly questionable; for example, a best-selling book on quality control shows a model rating scheme that is based in part on "percent of promises kept." Ratings are usually easy for vendors to manipulate by varying lot sizes or delivery dates or selectively sorting lots. Finally, and I believe most damning, rating schemes tend to draw attention from the true objective, never-ending improvement in quality.

PROCUREMENT STANDARDS AND SPECIFICATIONS

Quality control of purchased materials is an area that has been well explored. As a result, many of the basics of supplier quality control have been drafted into standards and specifications that describe many of the details common to most buyer and seller transactions. We will discuss some examples of commercial specifications as well as government specifications.

Most large companies have formalized at least some of their quality control requirements for purchased material. Traditionally, their specifications have covered the generic requirements of their supplier's quality system, including the details of any given purchase in the purchase order or contract. These supplier quality specifications typically cover such things as the seller's quality policy, quality planning, inspection and test requirements, quality manual, audits, the *supplier's* material control system, special process control, defective material control, design change control, and corrective action systems. The scope of the different specifications varies considerably from company to company, as does the level of detail. A good representative example is ANSI/ASQC Z1.15, Generic Guidelines for Quality Systems, available from the American Society for Quality Control.

The Department of Defense has formalized its quality system requirements in several different documents. The best known is Mil-Q-9858, which describes the general requirements for quality systems for prime contractors. Another standard, Mil-I-45208, describes the requirements for inspection systems. Mil-I-45208 is a subset of Mil-Q-9858.

SUMMARY

This chapter explained the importance of personal integrity in vendor quality assurance. The scope of vendor quality control was discussed. The separate and joint responsibilities of vendor and vendee were outlined. Some methods of evaluating vendor quality were introduced and their strengths and weaknesses described. Vendor rating systems were also covered. A few standards and specifications for procurement activities were mentioned.

This chapter did not describe detailed implementation of vendor quality control programs. Also omitted were the specific contents of procurement standards and specifications. The important topic of vendor quality audit was introduced but not covered in depth.

3

Human Resources and Quality

Most managers in every type of business discover the human element very early in their careers. Of course, no matter how hard we try, the ubiquitous human element usually remains. In fact, there are important functions performed *best* by people!

The situation is no different with quality control; human beings are a major source of quality problems. However, just as in the design activity, the quality system can be managed and designed to minimize the problems caused by human errors. Also, there are many important quality-related activities that humans still perform better than any machine. This chapter discusses the role of people in quality control and quality improvement.

BASIC PRINCIPLES OF INDUSTRIAL PSYCHOLOGY

There are several theories of human behavior vying for recognition. Although the science of psychology is over 200 years old, the effort to understand and explain human behavior is still in its infancy. Still, it has much to offer anyone interested in motivating people to do a better job.

Maslow's Hierarchy of Needs

Professor A. S. Maslow of Brandeis University has developed a theory of human motivation that has been elaborated on by Douglas Mc-Greggor. The theory describes a "hierarchy of needs." Figure 3.1 illustrates this concept.

Maslow postulates that the lower needs must be satisfied before one can be motivated at higher levels. Futhermore, as an individual moves up the hierarchy the motivational strategy must be modified because *a satisfied need is no longer a motivator.*

The hierarchy begins with physiological needs. At this level a person is seeking the simple physical necessitites of life, such as food, shelter, and clothing. A person whose basic physiological needs are unmet will not be motivated with appeals to personal pride. If you wish to motivate personnel at this level, provide monetary rewards such as bonuses for good quality. Other motivations include opportunities for additional work, promotions, or simple pay increases. As firms con-

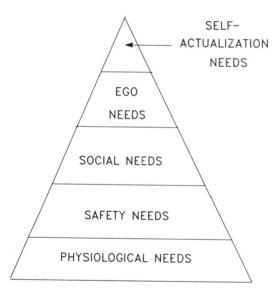

Figure 3.1 Hierarchy of Needs.

tinue to do more business in underdeveloped regions of the world, this category of worker will become more commonplace.

Once the simple physiological needs have been met, motivation tends to be based on safety. At this stage issues such as job security become important. Quality motivation of workers in this stage was once difficult. However, since the loss of millions of jobs to foreign competitors who offer better quality goods, it is easy for people to see the relationship between quality, sales, and jobs.

Social needs involve the need to consider oneself as an accepted member of a group. People who are at this level of the hierarchy will respond to group situations and work well on quality circles, employee involvement groups, or quality improvement teams.

The next level, ego needs, comprise needs for self-respect and the respect of others. People at this level are motivated by their own craftsmanship as well as by recognition of their achievements by others.

The highest level is that of self-actualization. People at this level are self-motivated. This type of person is characterized by creative self-expression. All you need do to "motivate" someone at this level is to provide an opportunity for him or her to make a contribution.

Herzberg's Hygiene Theory

Frederick Herzberg is generally given credit for a theory of motivation known as the hygiene theory. The underlying assumption of the hygiene theory is that job satisfaction and job dissatisfaction are not opposites. Satisfaction can be increased by paying attention to "satisfiers," and dissatisfaction can be reduced by dealing with "dissatisfiers." The theory is illustrated in Figure 3.2.

Theories X, Y, and Z

All people seem to seek a coherent set of beliefs that explain the world they see. The belief systems of managers were classified by McGreggor into two categories, which he called Theory X and Theory Y.

Under Theory X workers have no interest in work in general, including the quality of their work. Because civilization has mitigated the challenges of nature, modern humans have become lazy and soft. The job of managers is to deal with this by using "carrots and sticks." The carrot is monetary incentive, such as piece rate pay. The stick is docked pay for poor quality or missed production targets. Only money can motivate the lazy, disinterested worker.

Figure 3.2   Herzberg's hygiene theory.

Theory Y advocates believe that workers are internally motivated. They take satisfaction in their work and would like to perform at their best. Symptoms of indifference are a result of the modern workplace, which restricts what workers can do and separates them from the final results of their efforts. It is management's job to change the workplace so that the workers can, once again, recapture their pride of workmanship.

Theories X and Y have been around for decades. Much later, in the 1980s, Theory Z came into vogue. The Z organizations have consistent cultures where relationships are holistic, egalitarian, and based on trust. Since the goals of the organization are obvious to everyone and integrated into each person's belief system, self-direction is predominant. In the Z organization, Theories X and Y become irrelevant. Workers don't need the direction of Theory X management, nor does management need to work on removal of barriers, since there are none.

OPERATOR- VERSUS MANAGEMENT-CONTROLLABLE PROBLEMS

All the theories notwithstanding, a basic question that must be answered early is whether or not a particular class of problem is something we should expect the operator to be able to avoid. An operator-controllable problem has three distinct traits (Juran and Gryna, 1980):

1. The operators know what they are supposed to do.
2. The operators know what they are actually doing.
3. The operators have the responsibility, authority, skill, and tools necessary to correct the problems.

If any one of these traits is absent, then the problem is a management-controllable problem. Let's examine each of these three traits in more detail.

The Operators Know What They Are Supposed to Do

How often have you heard someone say, "if they would just do their jobs, everything would work out fine!" The number of lawsuits for breach of contract attests to the difficulty in clearly defining the requirements for any given assignment. Learning what the job is can be a difficult task.

Before operators can know what they are supposed to be doing, *management* must determine what the operators are supposed to be doing. This involves developing detailed procedures for the tasks that need to be performed and then training the operators in the proper interpretation of the procedures. The procedures must be written down; simply telling an operator what to do when he starts the job is not enough. It is also good practice to test the operator's understanding of the instructions. The test may be written, but verbal tests and demonstrations by actually doing the task under observation are also used.

Another aspect of knowing what is supposed to be done is the inspection standard, or the operational standard. Just how are "good" and "bad" defined? The definition should be as clear and unambiguous as possible, a task that is never as easy as it looks. For example, how would you define the printing standard for this book? You may find that such phrases as "the letters must be dark and crisp" mean different things to different people. Before an operator can know what he is supposed to do, such ambiguities must be dealt with. Until operator, inspector, supervisor, management, and end user all agree on the operational standard, the problem is a management problem, not an operator-controllable problem.

No company or institution operates solely by written procedures. Much of what is done involves *operating precedents*. An operating precedent is an established way of doing things. In all companies the operating precedent carries tremendous weight. If the written instructions contradict the operating precedents, the operator is placed in a very awkward position. Following the operating precedent means violating the written procedure, and following the written procedure means violating the operating precedent. For example, the written procedure might say to run a line at 5 to 7 units per hour, but it is traditional to run the line 10 units per hour at the end of the month when schedules get tight, even though the error rate tends to increase at the higher rate. If the operator has authority to change the line speed, he can be blamed for the additional errors if he sets the line to 10 units, but he'll also be the scapegoat for missed deliveries if he insists on holding to the written limits of 5 to 7 units. Because the resolution of such inconsistencies is beyond the control of the operator, the presence of any inconsistency between written practice and operating precedent make the problem a management problem, not an operator problem.

The Operators Know What They Are Actually Doing

Feedback is the essence of control. Unless the process output is evaluated, at least periodically, there is no basis for corrective action. Accurate timely feedback must be available to the operator if the problem is to be classified as an operator problem.

There are two basic types of process feedback data: variables measurements and attributes. Variables measurements are numbers like length, width, temperature, and weight. Attributes are characteristics possessed (or not possessed) by the process or the product. For example, a product may have a scratch, paint may be blistered, or a cleaning process may pass or fail a cleanliness test such as a water break test (if the water beads up the cleaning process fails, if the water sheets off the process passes).

If the operators are to use variables measurements to determine how they are doing, they must be given information on how many units to sample, how often to sample, what measurements to take, how to record the measurements, and perhaps what statistical process control analysis to perform on the data (see Chapter 7). The operators should be properly trained in the use of the measurement systems. If attribute data are to be taken, many of the same questions must still be answered; the operators must still know how many samples to check, how often to check the process, what type of statistical process control (SPC) analysis to use, etc.

Operators should get more than the immediate real-time feedback they need for process control. Trend reports and other long-term analyses should be made available to the operators. The analysis necessary for these reports is usually performed by the operators' supervisor or the quality control department, not the operators themselves. Also, if any additonal information is generated downstream, such as at the next operation or from the customer's experience with the output from the process, the operators should receive this information too. In other words, management should do all it can to assure that the operator gets complete information to use in process control and process improvement.

The Operators Have the Responsibility, Authority, Skill, and Tools Necessary to Correct the Problems

This means that when a problem occurs, the operators know what action is required to correct it and have the ability and authority to take the corrective action. Futhermore, if a problem occurs that an operator *can't* fix, he has the authority to take an alternative action, such as shutting down the process.

When action is taken, the operator must also have some feedback on the effectiveness of the action. This implies that the process reacts in a measurable way to the action. In most cases the feedback should come from the measurements in the "operators know what they are doing" section above. However, in some cases there must be faster feedback. For example, a manufacturer of ceramic parts has an operator quality control system. If the part quality indicates, the operator must adjust the temperature of a firing furnace. However, the temperature change must occur in a special zone within the furnace, and the change can be verified only with a special instrument normally kept in a laboratory. Since it takes several hours for a load of parts to run through the furnace, provisions are made for the operator to call for a special temperature check if he makes a temperature change.

Even though it is popular to blame the operator for most problems, the reader should know that most quality problems are *not* operator controllable. Deming (1986) estimates that 94% of all problems belong to the system (management). By applying the criteria described above, you will be able to separate problems into each category and thus assign responsibility properly. Chapter 7 discusses additional statistical criteria for making this determination.

CATEGORIES OF HUMAN ERRORS

When trying to eliminate or reduce errors, it is often helpful to "divide and conquer." By carefully dividing errors into different categories, we can sometimes better see what type of action is appropriate.

Inadvertent Errors

Many human errors occur because of lack of attention. People are notorious for their propensity to commit this type of error. Inadvertent errors have certain hallmarks:

There is usually no advance knowledge that an error is imminent.

The incidence of error is relatively small. That is, the task in normally performed without error.

The occurrence of errors is random, in a statistical sense.

Examples of inadvertent errors are not hard to find. This is the type of error we all make in everyday life when we find a mistake in balancing our checkbook, miss a turn on a frequently traveled route, dial a wrong number on the phone, or forget to pay a bill. These things can be overlooked at home, but not in quality control.

Preventing inadvertent errors may seem an impossible task. Indeed, these errors are among the most difficult of all to eliminate. And the closer the error rate gets to zero, the more difficult improvement becomes. Still, in most cases it is possible to make substantial improvements economically. At times it is even possible to eliminate the errors completely.

One way of dealing with inadvertent errors is *foolproofing*. Foolproofing involves changing the design of a process or product to make the commission of a particular human error impossible. For example, a company was experiencing a sporadic problem (note: the words "sporadic problem" should raise a flag in your mind that inadvertent human error is likely!) with circuit board defects. It seems that occasionally entire orders were lost because the circuit boards were drilled wrong. A study revealed that the problem occurred because the circuit boards could be mounted backward on an automatic drill unless the manufacturing procedure was followed carefully. Most of the time there was no problem, but as people became more experienced with the drills they sometimes got careless. The problem was solved by adding an extra hole in an unused area of the circuit board panel and then adding a pin to the drill fixture. If the board was mounted wrong, the pin wouldn't go through the hole. Result: no more orders lost.

Another method of reducing human errors is automation. People tend to commit more errors when working on dull, repetitive tasks. Also, people tend to make more errors when working in unpleasant environments, where the unpleasantness may arise from heat, odors, noise, or a variety of other factors. It happens that machines are very well suited to exactly this type of work. As time passes, more progress is being made with robotics. Robots redefine the word "repetitive." A highly complicated task for a normal machine becomes a simple repetitive task for a robot. Elimination of errors is one item in the justification of an investment in robots. On a more mundane level, using simpler types of automation such as numerically controlled machining centers often produces a reduction in human errors.

Another approach to the human error problem is ergonomics, or human factors engineering. Many errors can be prevented through the application of engineering principles to design of products, processes, and workplaces. By evaluating such things as seating, lighting, and sound levels, temperature change, and workstation layout, the environment can often be improved and errors reduced. Sometimes human factors engineering can be combined with automation to reduce errors. This involves automatic inspection and the use of alarms (lights, buzzers, etc.) that warn the operator when he's made an error. This approach is often considerably less expensive than full automation.

Technique Errors

I was once involved with a problem with gearbox housing. The housings were gray iron castings and the problem was cracks. Our supplier was made aware of the problem, and the supplier's metallurgist and engineering staff worked long and hard on the problem, but to no avail. Finally, in desperation, we sat down with the supplier to put together a "last-gasp" plan. If our plan failed, we simpy had to try an alternative source for the casting.

As you'd expect, the plan was grand. We identified many important variables in the product, process, and raw materials. Each variable was classified as either a "control variable," which we would take steps to hold constant, or an "experimental variable," which we would vary in a prescribed way. The results of the experiment were to be analyzed using all the muscle of a major mainframe statistical analysis package.

All of the members of the team were confident that no stone had been left unturned.

Shortly after the program began, I received a call from our quality engineering representative at the supplier's foundry. "We can continue with the experiment if you really want to," he said, "but I think we've identified the problem and it isn't on our list of variables." It seems that the engineer was in the inspection room inspecting castings for our project and he noticed a loud "clanging sound" in the next room. The clanging occurred only a few times each day, but the engineer soon noticed that the cracked castings came shortly after the clanging began. Finally, he investigated and found that the clanging sound was a relief operator pounding the casting with a hammer to remove the sand core. Sure enough, the cracked castings has all received the "hammer treatment"!

This example illustrates a category of human error different from the inadvertent errors described earlier. Technique errors share certain common features:

They are unintentional.
They are usually confined to a single characteristic (e.g., cracks) or class of characteristics.
They are often isolated to a few workers who consistently fail.

Solution of technique errors involves the same basic approaches as the solution of inadvertent errors, namely automation, foolproofing, and human factors engineering. In the meantime, unlike inadvertent errors, technique errors may be caused by a simple lack of understanding that can be corrected by developing better instructions and training.

Willful Errors (Sabotage)

This category of error is unlike either of the two previous categories. Willful errors are often very difficult to detect; however, they do bear certain trademarks:

They are not random.
They "don't make sense" from an engineering point of view.
They are difficult to detect.
Usually only a single worker is involved.
They begin at once.
They do not occur when an observer is present.

Perhaps an example would be helpful. An electromechanical assembly suddenly began to fail on some farm equipment. Examination of the failures revealed that the wire had been broken *inside the insulation*. However, the assemblies were checked 100% after the wire was installed and the open circuit should have been easily discovered. After a long and difficult investigation, no solution was found. However, the problem went away and never came back.

About a year later I was at a company party when a worker approached me. He said he knew the answer to the "broken wire mystery," as it had come to be known. The problem was caused, he said, when a newly hired probationary employee was given his two weeks notice. The employee decided to get even by sabotaging the product. He did this by carefully breaking the wires, but not the insulation, and then pushing the broken sections together so that the assembly would pass the test. However, in the field the break would eventually separate, resulting in failure. Later, I checked the manufacturing dates and found that every failed assembly had been made during the two weeks prior to our saboteur's termination date.

Often, the security specialist is far better equipped and trained to deal with this type of error than management, quality control or engineering personnel. In serious cases, criminal charges may be brought as a result of sabotage. If the product is being made on a government contract, federal agencies may be called in. Fortunately, willful errors are extremely rare. They should be considered a possibility only after all other explanations have been investigated and ruled out. These investigations are not wasted; they are necessary to prove sabotage.

SUMMARY

This chapter described the basic principles of industrial psychology that are especially relevant to quality control. The basic theories of management behavior, Theories X, Y, and Z, were discussed. Features that distinguish operator- and management-controllable problems from one another were described. A simple method of classifying and preventing human errors was presented.

The chapter did not discuss cultural and anthropological concerns, which may be very important for firms doing business in foreign countries or in disadvantaged areas of the United States. Also not covered were the myriad specific approaches to employee motivation such as employee involvement or quality circles. Human factors engineering methods, while introduced, were not described in detail.

4

Quality Organization and Management

This chapter describes the basic tasks required to assure that products and services meet the quality objectives of management. The emphasis is necessarily on tasks. However this presents a danger that must be understood very clearly: *Quality must be judged in terms of the final result, not in terms of completed tasks!*

The quality of your product or service, as judged by the end user, is the final criterion for measuring success. This simple lesson is absolutely critical to the long-term viability of the company. It is extremely easy for a company to inadvertently slip into the trap of judging quality by asking if everyone has met their responsibilities. That misses the point of the quality effort, namely to achieve quality judged by the consumer to be worth the cost relative to that of your competitors.

QUALITY POLICY

The first step in the quality management process is up to top management. What is the company's policy toward quality? While this may seem simple, it is the essence of the entire quality effort. The policy should be given a great deal of thought. Platitudes won't do ("XYZ corporation believes in quality"). The quality policy should be a guide to everyone in the company. If the policy is to provide competitive quality at a lower price, then the decisions of marketing, engineering, and production will be different from the decisions that would be made if the policy was to be the quality leader at a premium price. Everything from the advertising campaign to vendor selection to employee training depends on the policy of top management.

Policy statements are usually quite vague. This need not be the case. Although the company policy statement is not the place for detailed dissussions of responsiblity, it must be specific enough to provide the required guidance to everyone in the company. Ford Motor Company's mission statement is a case in point:

> Ford Motor Company is a worldwide leader in automotive and automotive related products and services, as well as in newer industries such as aerospace, communications, and financial services. Our mission is to improve continually our products and services to meet our customer's needs allowing us to prosper as a business and to provide a reasonable return for our stockholders, the owners of our business.

A number of things are made clear by this mission statement. First, the focus of the company is external. The position of leadership can only be measured in relation to the company's competitors. The customer's needs guide the company. The company's scope is defined. And Ford plans to provide a reasonable return consistent with its mission. This mission statement tells employees that the status quo will not be tolerated as a performance standard. They must seek ways to improve their products and services relative to the competition, and they must evaluate their success by looking to the customer. All of this must be done while providing a reasonable return on the investment of the stockholder, which is not necessarily the highest in the industry.

Perhaps the best way to illustrate the guidance provided by this particular mission statement is to contrast it with another one. Let's create a mythical company, "Econo Car Corporation." The policy of this company might state:

> Econo Car Corporation is a world leader in providing transportation to the cost-conscious consumer. Our mission is to provide safe, basic transportation at the lowest price.

The difference in basic philosophy is obvious. What is less apparent is the impact of this on day-to-day decisions. Nearly every action taken by every employee would be influenced to some degree by the policy. Everything from the paper used by the office typing pool to the target audience for the advertising would be affected. Clearly, the company policy statement should be given a great deal of thought. There should be nothing in the policy statement that is not endorsed by *every member* of the senior staff. Consensus is an absolute requirement.

COMPANY ORGANIZATION FOR QUALITY

In the preindustrial age the craftsman sold directly to the end user, whom he usually knew personally. With this arrangement there was no need for any kind of formal organization, for quality or any other reason. This began to change with the industrial revolution. Manufacture of interchangeable parts required at least a rudimentary inspection activity to assure that the various parts fit together. Factories were built in such a way that different activities were performed in different areas, such as the foundry, the machine shop, and assembly. Organizations tended to mimic the layout of the factory. In the early 1900s mass production methods were pioneered by Henry Ford and the management systems adjusted accordingly. "Scientific management" emerged as the dominant management style. Among other things, scientific management involves the division of labor and organization along the lines of the tasks managed.

Early organizaitons reflected the minor role of quality. Quality, when it was considered at all, was deemed to be the responsibility of the production department. When scientific management caught on, people with the title of "inspector" began to appear. An inspector was often the last person on the assembly line. The problem with this arrangement was that the inspector reported to the person who was responsible for production. Often, merchandise that was unfit for use was shipped anyhow.

The next step in this evolution was to create a separate department for inspection. The inspection department was usually headed by an inspection foreman who reported to the plant manager. In larger organizations, a chief inspector position was created. The chief inspector supervised the activities of several inspection foreman. The inspection-oriented quality organization persisted between World War I and World War II.

The existence of an independent inspection activity highlighted the cost of rejects. The obvious waste made many people ask if the defects couldn't be prevented from occurring in the first place. World War II dramatized the waste. Furthermore, techniques developed by Walter A. Shewhart of Bell Laboratories provided a scientific means of preventing defects through process control. Shewhart's methods of statistical process control, supplemented by other statistical methods developed during the 1930s and 1940s, became known collectively as statistical quality control, or SQC, methods. During the postwar period there was tremendous enthusiasm for SQC. The emphasis was clearly different from that of inspection. Also, the people who filled the positions were, typically, engineers. The job title of many of these new professionals was "quality control engineer." The organization structure was duly changed to accommodate the new positions in a separate department known as the quality engineering department. Quality engineering, along with the inspection department, reported to a new position, quality control manager.

The complexity of products and systems continued to increase following World War II. Many of the new systems failed in service from problems not related to the traditional arena of quality control, which was manufacturing. The failures were inherent in the design. It became necessary to broaden the scope of quality-related activities to include the design function. Quality planning became more widespread, and a new specialty, reliability engineering, made its appearance. Once again, the organization changed to accommodate these new activities.

The next major change involved increasing activism by the consumer. As consumers gained political strength, legislation was passed making effective quality control a requirement rather than an option. Regulatory bodies imposed standard requirements for quality control, such as the FDA's Good Manufacturing Practice (GMP). Other agencies were created to act as advocates for consumers, such as the Consumer Product Safety Commission. Agencies became much more aggressive on behalf of their consumer constituency and recalls of defective products became commonplace, costing manufacturers billions of dollars. Finally, changes in tort law made it easier for consumers to bring (and win) suits against manufacturers for placing defective products into the stream of commerce. All of these things acted to push the visibility of quality upward on the organization chart. By 1980 it was common to have vice-presidents of quality in larger organizations, and even quite small organizations had quality managers reporting directly to the chief executive officer. A typical organization chart is shown in Figure 4.1.

Work Elements

The work elements of quality control are divided among the organizational units in Figure 4.1 as shown in Table 4.1.

TASKS AND THE TOTAL QUALITY SYSTEM

The organization chart and discussion above are misleading in that they suggest that quality is the responsibility of specialists in the quality department. In fact, this is decidely *not* the case. As Chapter 1 emphasized, the specialists in the quality department have no more than a secondary responsibility for most of the important tasks that affect quality. The role of others in the company can be better understood if we look at a listing of the basic tasks as part of a system, which we will refer to as the total quality system (TQS). The TQS can be viewed as a process that assures continual improvement while implementing the policy established by top management. The TQS process is depicted in Figure 4.2.

The TQS makes it clear that the quality department shares the responsibility for quality with many others throughout the company. Let's examine Figure 4.2 more closely.

Figure 4.1 Quality Organization Chart

Table 4.1   Work Elements

Reliability engineering work elements
   Establishing reliability goals
   Reliability apportionment
   Stress analysis
   Identification of critical parts
   Failure mode, effects, and criticality analysis (FMECA)
   Reliability prediction
   Design review
   Supplier selection
   Control of reliability during manufacturing
   Reliability testing
   Failure reporting and corrective action system

QC engineering work elements
   Process capability analysis
   Quality planning
   Establishing quality standards
   Test equipment and gage design
   Quality troubleshooting
   Analysis of rejected or returned material
   Special studies (measurement error, etc.)

Quality assurance work elements
   Write quality procedures
   Maintain quality manual
   Perform quality audits
   Quality information systems
   Quality certification
   Training
   Quality cost systems

Inspection and test work elements
   In-process inspection and test
   Final product inspection and test
   Receiving inspection
   Maintenance of inspection records
   Gage calibration

Vendor quality control
   Preaward vendor surveys
   Vendor quality information systems
   Vendor surveillance
   Source inspection

QIS = Quality Information Systems.
MBWA = Management By Wandering Around.

Figure 4.2 Total Quality System (TQS)

TQS Policy

As discussed earlier, top management is responsible for establishing the company policy on quality. This policy forms the basis of the requirements.

TQS Requirements

Requirements are established for key products, processes, and vendors. Product requirements are based on market research regarding the customer's needs and the ways in which competitors' products meet or fail to meet these needs. Depending on the policy of the company, a product design must either compete with existing products or improve on them. Process requirements assure that designs are reproduced with a minimum of variability, where "minimum" is interpreted in a manner consistent with established policy. Process design is usually the responsibility of process engineering specialits. Vendor selection is typically based on reliability, price, and quality. Purchasing bears the primary responsiblity, with quality control and engineering providing assistance.

TQS Measurement Systems

The term "measurement system" is used here in the broadest sense. The purpose of a measurement system is to determine the degree to which a product, service, vendor, process, or management system conforms to the requirements established. At times the measurement is not straightforward. For example, design conformance is accomplished by a design review. Quality control, marketing, and manufacturing evaluate the design to determine whether it meets the customer's expectations and can be economically and consistently reproduced by manufacturing. This is a subjective evaluation, but the parties involved should reach a consensus.

Quality cost measurement is such an important subject that it is discussed in depth later in this chapter. Suffice it to say that the design of quality cost systems is usually the joint responsibility of the accounting and quality control departments.

Product and process measurements are designed to assure that the products and processes conform to requirements. This usually involves obtaining one or more observations and comparing the observations to operational standards. Although the quality control department may be responsible for obtaining measurements on certain products and processes, the responsibility for acting on the results remains with the operating department.

Measurement of service quality has been woefully inadequate. In fact, many companies fail to reconize their responsibility to the customer in this area. In most cases of neglect it takes an external force, such as a government agency or a competitor, to create a change. Unlike the situation in most of the other areas, responsibility for measuring service quality has not been established by precedent. However, it seems logical that quality control and marketing share a joint responsibility.

Measurement of vendor quality was discussed in Chapter 2.

Quality control and purchasing were the primary players in developing the measurement systems. Of course, the final responsibility for quality remained with the vendors themselves.

Special studies are different from the other measurement systems described here in that they are not an ongoing activity. Studies may be commissioned to answer an important question or to obtain vital facts. For example, a competitor may have just released a revolutionary product that requires close examination. The responsibility for special studies is usually stipulated in the charter that commissions the study.

Finally, remember that as stated early in this chapter, the proper emphasis is on results, not task completion. The most important measurement is therefore, customer satisfaction. Marketing is usually given responsibility for developing systems that measure customer satisfaction.

TQS Audit

The purpose of a TQS audit is to seek verification that an adequate system exists and is being adhered to. The primary responsibility for auditing TQS is top management's. In addition, the manager of each activity must perform audits within his or her area to assure that the requirements are being met. Finally, "third party" audits by the quality department or auditors from outside the company are also common. All third party audits should be commissioned by management. However, the third party audit should never be the primary source of detailed information for management. There is no substitute for firsthand observation.

Basic ground rules must be established before proceeding with an audit. First, there must actually be a requirement against which performance is audited. The performance standard should be written and communicated to the responsible parties. Second, there must be an established means of measuring performance. Finally, the responsibility for action must be unambiguous and clearly understood. Third party audits should be preannounced whenever possible.

Audits can be divided into two basic categories. The *quantitative audit* is designed to provide a score that indicates the degree of compliance of a program or quality system. The *qualitative audit* is a generalized assessment of the effectiveness of a quality program. Since in practice a quality program has elements that are both quantitative and qualitative, it is rare to find any audit that is purely qualitative or quantitative; most are a mixture of both. In general, the qualitative audit is less formal, and it is the type of audit that should be conducted by internal personnel, including management, auditing their own operations. The quantitative audit is normally conducted by third parties, and it provides an objective assessment that can be used to spot trends in performance.

Audits serve many different purposes. At a financial level they can provide a means of evaluating the effectiveness of quality cost expenditures. Properly done, audits can provide a means of motivating people to higher levels of performance. Audit results can provide a guide to action by identifying system elements that might be the cause of problems. Audits can be used as an early warning system that indicates potential trouble before it actually leads to a negative customer reaction.

Audits can also be classified according to just what is being audited. At the broadest level, the *quality program audit* assesses the appropriateness of the quality program in relation to the company policy and mission. Quality programs range from minimal to extremely elaborate, and it may be that the formal quality program does not fit the company's objectives or industry. Given that the quality program accurately reflects the wishes of top management regarding scope and size, the *quality system audit* can be conducted. This audit measures the overall effectiveness of the quality program and provides guidance for improving the effectiveness. *Audits of procedures* are conducted to determine the extent of compliance to procedures. At times such audits uncover procedures that are outdated or ineffective or show a need for formal procedures where none currently exist. Also common are *product audits*. Product audits can take place either while the product is in process or on the finished product. Product audits are usually in-depth evaluations of very small samples to determine the extent of compliance to established product requirements. The product auditor should perform the evaluation from the viewpoint of the end user or, at times, of the outside regulator. *Process audits* are conducted to evaluate the conformance of materials and processes to established requirements. Statistical process control (SPC) by operators is a type of process audit, as is the periodic review of control charts by area managers. Finally, *special audits* are often commissioned by management to get needed facts on the company's operations. These audits can range from opinion surveys to housekeeping checks, or any other subject that interests management.

TQS Feedback

All of the information collection in the world is absolutely useless without an effective feedback mechanism. The feedback loop of the total quality system is closed by a variety of means. The most visible is the formal quality information system (QIS) of the company. The QIS encompasses all of the formal, routine activities associated with quality data collection, analysis, and dissemination. The subject is a big one and is covered in detail in Chapter 5.

Formal systems, while necessary, provide only part of the feedback. Human beings may get the facts from these systems, but they can't get a *feeling* for the situation from them. This level of feedback comes only from firsthand observation. One company encourages its senior managers, in fact *all* of its managers, to "wander around" the work areas to observe and to interact with the people doing the work. The term given to this activity is management by wandering around, or MBWA. I can personally attest to the effectiveness of this approach, having done it myself and seen it work. There are many "facts" that never appear in any formal report. Knowing that quality costs are 10% of sales is not the same as looking at row upon row of inspectors, or a room full of parts on hold for rework, or a railroad car of scrap pulling away from the dock.

Just as MBWA provides direct feedback from the workplace, a company needs direct personal input from its customers and suppliers. Company personnel should be encouraged to visit customers. All managers should be required periodically to answer customer complaints (after being properly trained, of course). A stint on the customer complaint "hot line" will often provide insight that can't be obtained in any other way. Vendors should be asked to visit your facility. More formal methods of getting customer and vendor input are also important. Vendors should be made part of the process from design review to postsale follow-up. Often a slight design modification can make a part much easier to produce, and the vendor, who knows his processes better than you do, will often suggest this modification if he's invited to do so. Customer input should be solicited through opinion surveys, customer review panels, and the like. Of course, the QIS will provide some of this input.

Improvement

The whole point of getting feedback is improvement. Why bother collecting, analyzing, and distributing information if it will not be used? This simple fact is often overlooked, and many companies devote vast resources to developing ever more sophisticated management information systems that produce little or no improvement. Discovering this situation is one of the primary objectives of the quality audits discussed earlier.

Continuous improvement should be a routine part of everyone's job, but making this actually happen is extremely difficult. Most operating precedents and formal procedures are designed to maintain the status quo. Systems are established to detect negative departures from the status quo and react to them. Continuous improvement implies that we constantly attempt to change the status quo for the better. Doing this wisely requires an understanding of the nature of cause systems. Systems will always exhibit variable levels of performance, but the nature of the variation provides the key to the type of action that is appropriate. If a system is "in control" in a statistical sense, then all of the observed variability is from common causes of variation that are inherent in the system itself. Improving performance when this situation exists calls for fundamental changes to the system. Other systems will exhibit variability that is clearly nonrandom in nature. Variation of this sort is said to be due to "special causes" of variation. When this is true it is usually best to identify the special cause rather than to take any action to change the system itself. Changing the system when variability is from special causes is usually counterproductive, as is looking for "the problem" when the variability is from common causes. Determining whether variability is from special causes or common causes requires an understanding of statistical methods. Some of the more useful and easily understood methods are described in subsequent chapters of this book.

Quality improvement must be companywide in scope. The entire cycle from marketing and design through installation and service should be geared toward improvement. This holds regardless of the policy of the company. Companies that aspire to be cost leaders have as much need for improvement within their market segment as those that aim for quality leadership. A key part of this effort must be continuous improvement of the management systems themselves.

THE QUALITY MANUAL

Formal quality systems are documented in the quality manual. The manual contains company policies and procedures that affect product quality. Quality audits by third parties often begin with the quality manual. Most quality manuals follow a similar format. For example, the table of contents shown in Table 4.2 is from the *Manual of Quality*

Table 4.2 Contents of a Quality Manual

Introduction.

Administration.
    Quality Procedures and Bulletins.
    Stamp and signature administration.
    Audit and surveillance administration.
    Training and certification administration.

Quality program management.
    Organization and functions.
    Quality planning.
    Work instructions.
    Records and reports.
    Corrective action.
    Costs related to quality.

Facilities and standards.
    Drawings, documentation and changes.
    Measuring and test equipment.
    Production tooling as media of inspection.

Control of purchases.
    Candidate supplier/subcontractor quality evaluation and selection.
    Active supplier/subcontractor quality compliance.

Manufacturing control.
    Materials and material control.
    Production processing and fabrication.
    Completed item inspection and testing.
    Handling, storage and delivery.
    Nonconforming material.
    Statistical quality control and analysis.

Coordinated customer/contractor actions.
    Customer inspection at subcontractor or vendor facilities.
    Customer property.

Assurance Procedures and Forms (Carlson et al., 1981), which is based on the requirements of Mil-Q-9858A, a military standard for the quality programs of prime contractors. Quality manuals must be updated constantly. Typically, manuals are serialized and those with registered manuals are sent bulletins as changes are made.

In my work I have seen dozens of quality manuals. Subsequent audits of the systems documented by these manuals lead me to believe that the size of the manual has no relationship to the effectiveness of the quality system. Good quality manuals, like the effective systems they document, are well thought out and customized to the particular needs of the company. The emphasis is on clarity and common sense, not on an effort to document every conceivable eventuality. For example, the quality manual of a very large defense contractor listed *68 ways* to identify scrap material (red paint, stamped with an "R," a scrap tag, etc.). This approach led to constant difficulty because there were always new cases not allowed for in detail. In one instance a government auditor complained because some defective microchips were not properly identified. The chips were in a plastic bag, and even though a properly completed scrap tag was in the bag with them, the parts themselves were not stamped with an "R" (it's hard to stamp a part only a fraction of an inch square!). As you might expect, the company quality control manual was several hundred pages thick and very few people outside the audit department had ever read it. In contrast, I audited a medium-sized electronics company whose quality manual consisted of 45 well-written pages. The scrap identification problem was handled by a single sentence that stated: "Scrap material will be clearly and permanently identified." In my audit of the quality systems I found that all registered manual holders had read the manual and understood it. The president of the company was a registered manual holder.

QUALITY COSTS

The American Society for Quality Control (ASQC) divides quality costs into four categories:

Prevention costs: Costs incurred to prevent the occurrence of non-conformances in the future

Appraisal costs: Costs incurred in measuring and controlling current production to assure conformance to requirements

Internal failure costs: Costs generated before a product is shipped as a result of nonconformance to requirements

External failure costs: Costs generated after a product is shipped as a result of nonconformance to requirements

These cost categories allow the use of quality cost data for a variety of purposes. Quality costs can be used for measurement of progress, for analysis of problems, or for budgeting. By analyzing the relative size of the cost categories, the company can determine if its resources are properly allocated.

As stated earlier, the quality cost system should be developed jointly by the quality and accounting departments. Although the task of monitoring quality costs on an ongoing basis is normally the responsibility of the accounting department, in some companies the quality department takes the lead. Because of this, the quality cost system is often thought of as an accounting system or a quality department system. This is a mistake. The quality cost system is primarily a *management information system* whose purpose is to guide management in their efforts to improve quality. The point of making this distinction is that one should not go to extremes in trying to get to-the-penny accuracy. By their nature, quality costs are often difficult to determine, and at times a good approximation is the best you will be able to obtain. Usually, this is all you need to determine the appropriate management action. The practice of overemphasizing to-the-penny accuracy is a common cause of failure of quality cost programs (Juran and Gryna, 1980, pp. 30–31).

Quality cost elements, while sometimes difficult to measure, are not difficult to categorize. Table 4.3 provides a guide to some of the more commonly encountered quality cost elements.

Quality costs are an extremely important means of directing management action. They can also be very useful in helping management track the success of their quality improvement efforts. Ideally, the total cost of quality (COQ) will decline over time. Crosby (1979) recommends a 10% per year goal for reducing total COQ, but this figure seems somewhat arbitrary. A better approach would be to make a realistic assessment of the *causes* of current quality cost elements and develop an improvement plan based on this assessment. This approach requires an in-depth understanding of quality technology by upper management and the use of SPC and other statistical methods introduced in this book. One especially helpful quality tool is the quality information system (see Chapter 5), which should provide the detailed data on which the COQ figures are based.

Table 4.3 Quality Cost Elements

Prevention costs
  quality planning
  process control planning
  design review
  quality training
  gage design
Appraisal costs
  receiving inspection
  laboratory acceptance testing
  in-process inspection
  quality audits
  outside endorsements (e.g. Laboratory approval)
  calibration
  inspection and test equipment
  field testing
Internal failure costs
  scrap
  rework
  process troubleshooting
  vendor caused scrap or rework
  material review board activity
  re-inspection or re-test
  downgrading
External failure costs
  processing of customer complaints
  service
  unplanned field repair
  recalls
  processing of returned materials
  warranty

IMPACT OF QUALITY COSTS ON PROFITABILITY*

The basic factors of production are land, labor, capital, and (some economists include) management. Note that these factors also con-

---

stitute the source of all cost. Thus most managers equate cost to output and assume that the two are inextricably related. If we call costs "inputs" and the production that results from the inputs "output," then the model of production in the minds of most managers is

Increased input = increased output

An analogous mental model of profitability is that it is inextricably linked to sales volume. This model can be described with the equation

Increased sales = increased profit

From these two equations it seems to follow that the most direct means available for increasing the absolute level of profits is to increase the absolute level of sales. As we shall see, in many cases this assumption is absolutely untrue.

An alternative means of increasing profit is to reduce costs. Cost reduction effected through a given program often seems trivial when expressed as a percentage of total sales. However, the direct impact of cost reduction is *not* on sales. Obviously, cost reductions have their most direct impact in the area of costs (by definition). Also note that costs are the converse of profits. The sum of costs and profits equals sales (in a poorly managed company, the sum of costs and losses).

Since the ratio of profit to sales is often large, it is important that the impact of a cost reduction be stated in a way that accurately reflects its true importance, which is, to repeat, in the area of cost reduction and profitability and *not* in the area of sales. It was stated that sales equals costs plus profits. The usual justification for an increase in costs (investment) is that it will earn a return. Generally, this means an increase in the profitabiliy figure that will someday exceed the adverse impact on costs. Usually this assumes an increase in sales. The logic is that profit is a function of sales. The common phenomenon of early loss on new investment is revealed in the sales formula restated to express the profit relationship:

Profit = sales − costs

The expected increase in sales suffers only a production lag but the increase in profits expected from a new investment always suffers a cost lag, frequently called a "payoff period."

Given this normal relationship, we can analyze the effect of a hypothetical investment on sales, costs, and profits. Let us assume that our investment will yield a return of 20% and that it will be externally funded and not taken from current profits. We assume that prior to our investment our firm's financial statement looked like this:

Sales $10,000
Cost   9,500
Profit   500

We further assume that the new investment will only expand the capacity of our firm and will not affect the percentage of return on sales. This means that we will continue to have profitability of 5% return on sales as shown above even after we invest in the additional capacity. Given these assumptions, here is what the financial statement would look like after we invested the $500:

Sales $12,000
Cost   11,400
Profit   600

You can see that the increase in profit of $100 constitutes a 20% return on the $500 investment. If we are to maintain our 5% return on sales, this profitability must come from an additional $2000 in sales. Since cost = sales − profit, the cost of operating the new capacity must be $11,400. This is an additional cost of $1900, more than 2½ times the $500 investment! It is the operations costs that offer great potential for improved profitability.

In the foregoing example we attempted to increase profit by increasing sales. Note, however, that in the example the sales figure is 20 times the profit figure. Also, the relationship of sales, profit, and cost is linear. Given this linear relationship, it is apparent that cost changes that do not affect the sales figure will alter the profit figure by an amount equal to the change in costs. Most costs have a beneficial effect on sales, but there is one category that does not: failure costs. This fact is significant in that failure cost reductions become, in effect, technological improvements, and they enable firms to overcome the adverse impact on profits caused by diminishing returns to scale.

When one input is variable, the relationship between input and product is conventionally divided into three stages, as illustrated in Figure 4.3.

Figure 4.3   Stages in the life of a firm.

In stage 1 the average product per input increases. For example, two workers would produce more than double the output of one worker. The point where marginal and average products are equal and average product is at a maximum is the boundary of stage 1.

In stage 2, total product continues to increase but at a decreasing rate. The right-hand boundary of stage 2 is at maximum total product and zero marginal product. In stage 2 also, both average and marginal product are on the decline; marginal product, being below the average product, pulls the average down.

In stage 3, total product is declining. No rational producer operates for any period of time in stage 3. A firm that is near the boundary of stage 3 has little to gain from an increase in sales. Any sales increase will require increased output to meet it, and this will move the producer even closer to stage 3 and negative marginal product. The producer would actually suffer a *loss* from increased sales.

When a firm is in a market dominated by a few sellers (oligopoly), increased output from new capacity presents a different problem. In this case unsatisfied demand must exist prior to the increase in output since it cannot be created via price cuts. In oligopolistic markets the competitors would simply follow suit, and the result would be to *increase* the cost figure while *decreasing* the dollar volume of sales. Thus, new investment in a market characterized by oligopoly is predicated on the assumption that demand for the product will increase given the current price.

Obviously, a beneficial impact on profit in the foregoing situations cannot be best realized by increasing output on sales. The input variable required is technology, which alters the relationship between factor inputs and marginal productivity. In looking at Figure 4.3, we see that a favorable impact on marginal productivity can move a firm *backward* on the growth curve, from a stable or even declining growth curve to a position of more dynamic growth, including growth in profit.

Consider some reasons for this phenomenon. The level of profit is increased in both relative and absolute terms; more investment is attracted because of the higher returns; consumer demand that went unsatisfied because of the negative impact of diminishing returns to scale can now be met; firms improve their positions relative to their competitiors; and so on, ad infinitum.

Failure Costs

Two categories of investment produce the beneficial effect described above: R & D investment and investment in the reduction of failure costs. We are, of course, concerned with the latter category. Reduction of failure costs has a direct impact on the bottom line because *it has no adverse effect on output*. Like an R & D investment, reduction of failure costs produces new technology in the sense that more output is obtained from less input. Usually an equivalent impact from return on sales requires a much greater investment, and, as we have seen, often sales dollars are unable to make *any* profit contribution.

In economics the term *opportunity cost* is defined as the dollar amount that would be derived from the employment of a factor of production in the best alternative use. In the purest sense, failure costs are opportunity costs, lost units of output from factors of production. And the opportunity foregone is profit.

Management deserves to be informed of these opportunity costs and of their true importance. Management is used to equating profit to sales. It can be an eye-opener if, in addition to the customary measurement bases, management is told what volume of sales it would take to generate as much profit as is foregone in failure (waste) costs.

The figures show the potential improvement in profitability. Realizing this potential requires an investment in quality technology. Investing in higher profit through improved ability to eliminate waste is every bit as sensible as investing in additional capacity and sales. In many cases, it is the only sensible way to improve competitiveness.

SUMMARY

The importance of developing a consistent and meaningful quality policy was described. The evolution of the structure of quality organizations over time was covered. Work elements of the various groups in the typical quality organization were presented. Also covered were organizations for implementing the TQS process. Quality audit systems were described along with feedback systems. Quality manuals were introduced and their contents examined. Quality cost measurement systems were described, and the impact of quality costs on profitability was evaluated.

The chapter did not discuss industry-specific organizations for quality or the many different ways in which quality responsibilities are allocated in different companies.

5

Quality Information Systems

A quality information system (QIS) is used in collecting, analyzing, reporting, or storing information for the total quality system. This chapter discusses the fundamental principles of QIS, followed by a case study of one company's experience in establishing a QIS from scratch.

The QIS contains both manual and computerized elements and requires inputs from many different functions within the company. In many companies the QIS is one part of an even more cmprehensive system known as the management information system or MIS. The MIS is an effort to bring the data collection activities of the entire organization into a single integrated system. When it works properly, the MIS has a number of advantages. For one thing, it eliminates the duplication of effort that would be required if the MIS did not exist. Also, it is a globally optimized system (i.e., the *whole system* is designed to be of maximum benefit to the company). Without an MIS the information system is likely to be a collection of optimized subsystems that are "best" for the individual departments they serve but are probably not optimal for the company as a whole.

Management information systems also have their darker side. Even the concept itself can be questioned. "Dictatorship of data processing" is one common complaint. In designing the one grand system, data processing has the ultimate say in whose needs are served and when they will be served. Second, MISs are often so large and complex that changing them is nearly impossible. Finally; MISs are concerned almost completely with computerized systems; this makes them incomplete, since much vital information never makes it into a computerized data base.

PLANNING A QIS

Because a QIS requires inputs from a wide variety of different functions within the company, planning the system will involve a large number of people from these different groups. The quality department should not develop a QIS in a vacuum and expect it to be accepted. The QIS is not a quality department system, it is a *companywide system*. However, the quality department usually bears the major responsibilty for operating the system.

Although QIS means "computerized" to most people it is important to note that most successful QIS applications begin as successful *manual* applications. Usually a system element important enough to be made a part of the computerized QIS is too important to wait until it can be programmed; thus some form of manual system is established at least as a stopgap measure. Planning the manual system is no trivial task; data collection needs must be established, personnel selected and trained, reports designed and their distribution lists established, and so forth. Because of this vested work, manual systems provide a great deal of information that can be used in planning the more elaborate computer-based sytem.

The flowchart is a tool that has been used in planning computer systems since the beginning of the computer age. Flowcharts can be applied at all stages in the process, from system design to writing the actual computer program code. The standard symbols used in creating information system flowcharts are established by the American National Standards Institute (ANSI) and are shown in Figure 5.1.

A general approach to QIS planning is shown in Figure 5.2. Good QIS planning always starts with the benefits of the system. This drives the rest of the process. When possible, benefits should be expressed in dollars. However, the most important benefits are sometimes the most

Figure 5.1 Flowchart symbols.

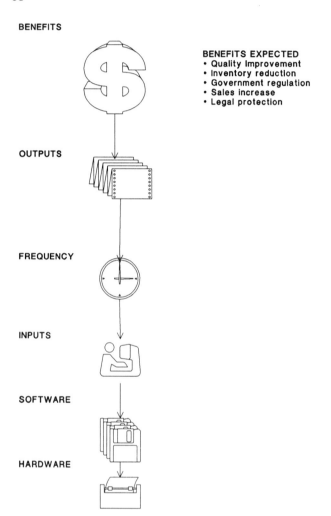

BENEFITS

BENEFITS EXPECTED
- Quality Improvement
- Inventory reduction
- Government regulation
- Sales increase
- Legal protection

OUTPUTS

FREQUENCY

INPUTS

SOFTWARE

HARDWARE

Figure 5.2   Planning flowchart for a QIS.

difficult to quantify, such as safety. Other QIS elements are required by law—for example, FDA traceability requirements.

After deciding that the benefits justify the creation of the QIS element, the next step is to define the outputs required. This should be done in as much detail as possible, as it will be useful in later stages of the planning process. Output requirements should be classified into "must have" and "would like to have" categories. Often the cost of a system is increased dramatically because of the effort required to add some optional item. Data processing professionals refer to these options as "bells and whistles." The form of the output—on-line data, periodic reports, graphics, etc.—can have a significant impact on the total cost of the system.

How often the output needed is also important. Continuous output in the form of a real-time system is at one extreme, while annual reports are at the other.

INPUTS

Inputs, including personnel and equipment requirements, can be determined at this point. The resources required to implement the system can be substantial.

SOFTWARE AND HARDWARE

Figure 5.3 illustrates the elements of a basic information system. The computer can receive information in a variety of ways. The methods can be divided into two categories: on-line systems and batch systems. On-line systems enter data directly into the computer data base (in real time), while batch systems go through an intermediate stage, such as keypunch, and the data base is updated periodically. Which system to use depends on the use to be made of the data. An inventory tracking system should be an on-line system since people will be using it to locate parts in real time. A quality cost reporting system could be a batch system since it must be up to date only when the monthly report is run. In general, batch systems are less expensive.

With on-line systems, the data are entered directly from the data collection point into the data base. These systems may involve typing in the information at an on-line terminal, reading it in via a bar code system, voice data entry, or input directly from automatic sensors without involving a human being at all. The technology is constantly changing,

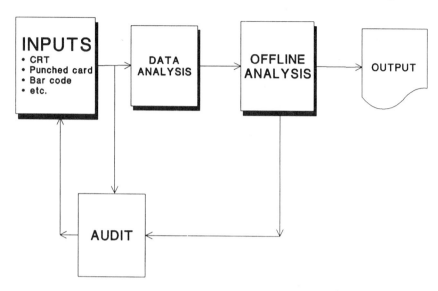

Figure 5.3 Elements of an information system.

and it is beyond the scope of this book to cover the subject completely. In selecting the appropriate technology, the system analyst must consider the environment in which the equipment will be used, costs, the amount of data to be collected, and the level of training and number of trained personnel required. Weighing the costs and benefits of the various technologies can be difficult.

Batch systems are generally more paper oriented. Manual systems being converted to computers often go through a batch processing phase on the way to becoming a real-time system. Because most batch systems work from a physical document, the training is directed at first getting the correct information on the document and then getting the information from the document into the computer. The existence of a physical document provides a built-in audit trail and greatly simplifies the task of verifying the accuracy of the information. In the past, transfer of the information from paper into the computer could be done only by having a keypunch operator create a punched card by typing the information all over again. However, new technology includes "smart" optical character readers (OCRs) that can convert information from typed, printed, or even handwritten documents. Computers can also tie in to other computers over telephone lines or through direct links, and radio transmission of computer signals is also possible. As with real-time data entry, the pace of change requires that you research the technology at the time of your project.

More important than the technology is the *accuracy* of the input. The computer trade coined the phrase "garbage in garbage out," or GIGO, to describe the common problem of electronically analyzing inaccurate input data. Accuracy can be improved in a variety of ways. Choosing the appropriate input technology is one way. Another is to have computer software check inputs for validity, rejecting obvious errors as soon as they are entered. However, no matter how much input checking takes place, it is usually necessary to have someone audit the input data to ensure accuracy.

DATA PROCESSING

When you are confident that the input data are accurate, you are ready to have the computer analyze the data. Whether to do the analysis on a microcomputer, a minicomputer, a mainframe computer, or a supercomputer is a question of balancing the need for speed and accuracy against the cost. You must also consider the need for other information stored elsewhere in the system, such as the need to access the part number data base. Again, the pace of change makes it essential that you review this question at the time you are designing your system. The mainframe of yesterday is on everyone's desk today.

OFF-LINE ANALYSIS

A truism of quality is that you can never get all of the defects out of the system. A parallel truism of data processing is that "all software has bugs." The result of these two things in combination is the production of computer reports that contain ridiculous errors. Thus, before distribution of any computer report, you should look it over for glaring mistakes. Is the scrap total larger than the sales of the company for the year? Are people who left the company last year being listed as responsible for current activities? Are known problem parts not listed in the top problems? This type of review can save you much embarrassment and can also protect the integrity of the information system you worked so hard to create.

REPORTING

Reports should be sent only to those who need them, and people should be sent only what they need. These simple guidelines are often ignored, creating an information overload for management. As a rule, the level of detail required in a report is greatest at the lower levels of the organization and least at the top. A storeroom clerk may need a complete printout of all the items in the storeroom, breaking out the number of parts in each location. The company president needs to know only the total investment in inventory and the trend. It is recommended that charts be used to condense and summarize large quantities of tabular data when possible.

The distribution of reports should be carefully monitored. From time to time, report recipients should be asked if they still need the report. Also, the report recipients are your *customers*, and their input regarding changes to the reports should be actively solicited.

All of the principles above come together when a QIS is designed and implemented. Few people have the opportunity to create a QIS from scratch. I was fortunate to have had the opportunity to do so. Although I was but one of many players, the experience was quite valuable.

The company involved was growing at a fantastic rate. Within a period of 5 years their employment skyrocketed from 350 people to 1600. As the company grew, their old systems became obsolete. Because of the rapid growth, most of their systems collapsed at the same time. This situation was, of course, a crisis for the company. However, it was also an opportunity to build new systems from the ground up. The case study documents the process of designing these systems from the perspective of a member of quality management.

CASE STUDY *

Not long ago our quality assurance (QA) program used *no* computer time, absolutely none. Yet our manufacturing business was skyrocketing toward $100 million and our product was becoming increasingly complex. Today we utilize computers intensively, including large in-house mainframe computers, time-sharing, and a microcomputer system. We have the ability to determine reject rates for any part, machine, workstation, department, or supplier. We can determine instantly whether a part is on hold, the quantity on hold, what is wrong, what is to be done, who is to do it, and exactly where the parts are in any of our 20-odd storerooms with thousands of individual locations. We know scrap rates on a part-by-part or department-by-department basis.

We have a detailed history on field problems, and in some special cases, we have a complete history by unit on high-problem parts (e.g., life history from assembly line to scrap pile). We can determine instantly which parts have caused the greatest field problems in the past and which are the current "top problems" as measured by a variety of means (dollars, most often claimed, greatest quantity claimed). We can call up on-line current warranty problems for any part and get detailed descriptions of problems encountered.

Each month, management receives the latest tally of warranty and field repair charges broken out by model, summarized, plotted, and measured in terms of dollars per unit, claims per unit, and percent of net sales billed.

Complex statistics are routinely computed on a time-sharing computer or a microcomputer. Sample sizes are kept to a minimum by using the most efficient (as opposed to the simplest) statistical methods. When the computer is the mathematician, one can scarcely notice the difference between simple and complex solutions anyhow, especially on a microcomputer when "coretime" is free. The computer's ability to do flawless manipulations of numbers is also useful in such error-prone operations as sorting. The ability to plot and create histograms makes "picturing" data distributions a snap.

We did not just arrive at where we are; we had to get here. Getting here involved considerable trial and error, interfacing with large numbers of people in many areas of the company, familiarizing ourselves with the new world of the computer. The process (which is far from complete) involved a lot of adjusting and some pain. People must be trained, programs must be debugged, procedures must be developed, implemented, and revised.

We found that the successful computer system is normally preceded by a successful manual counterpart. A manual system can be designed in a way that makes the transition to the computer less painful. At times, because of resource limitations, only part of a given system can be computerized, which can cause considerable difficulty if not properly planned and handled.

The computer's role in modern quality assurance is, like the computer itself, complex and often indispensiable. Yet the journey to the computer is fraught with danger and should not be begun without due appreciation of the pitfalls likely to be encountered along the way. By describing one company's rather typical trip down the road to computers, we hope to cast some light on a number of these pitfalls and make the trip easier for future travelers—and all signs indicate that the number traveling this road is likely to increase greatly in the near future.

I wish I could tell you that in the "early days" of my introduction to computers I could have foreseen that today, some 5 years later, we would be where we are now. But it "just ain't so." Back then our QC program was embryonic, having just grown from three people to six (I was number six). Our biggest problem was learning to communicate with one another from department to department and shift to shift. It has been my experience that most action is motivated by the need (actual or imagined) to solve a problem. For 2 years, as our company and department exploded with growth, our problems seemed distant from the computer. Questions such as "How should we organize?" and "Who should I hire?" and "What authority should be delegated?" did not lend themselves to electronic analysis.

Slowly, things began to change. Even though the rapid growth continued, there was a steady flow of data from the 20 people already employed in quality assurance and the need was felt to glean maximum benefit from these data. Thus was born our first "scrap report," a *manual* compilation of all parts scrapped in a given month. A first, the dollars were calculated by searching a computer listing for the cost per piece and then multiplying this by the number of pieces scrapped. This duty was first performed by QA technicians with the help of a part-time clerk, but it was soon obvious that the technicians were spending far too much time in this clerical task. Added to this was the complaint that the cost figures were inaccurate because the parts were scrapped at full cost, even though many were actually scrapped several operations before completion.

With these concerns in mind, the advertisement for a "canned" computer program that generated scrap reports fairly jumped from the page! There was our salvation! Just as a formality, we sent a copy of our letter requesting information to our data processing department (previous approaches to them had gotten us forms to fill out, steeped in the forbidding jargon of the computer world). Much to our surprise, an in-house systems analyst appeared at the QA office a short time later holding a copy of our letter. Why, he asked, did we want to take such a simple problem to an "outsider"? With extremely little resistance, we agreed to "allow" the scrap report to be run by our data processing department.

It seemed to me then (and still does) that people take more notice of a computerized report than of a manual one. So it was with our spanking new scrap report, and when the managers received the report, they promptly set about devising plans to reduce the costs reported. The early result of this attention was to point out a number of errors and shortcomings in the report itself. For a time, these had to be worked out by manually correcting the computer printout.

This was also the first year we began our program of quality engineering in earnest, and my new duties included the application of statistical analysis to day-to-day problems. Most of the analysis was straightforward and could be handled by calculator. Often I resorted to prepared forms, scatter plots, histograms, and the like. Still, inevitably a problem would come up where only a more advanced technique was suitable. Usually, I would first try to deal with the problem using simpler methods, thus wasting time. However, I would finally be forced to yield to the overwhelming logic of the situation and apply the more advanced techniques manually.

If you have never tried it, you should try sometime to work out a multiple regression fitting two or more independent variables by the least-squares method using only a pocket calculator. You will find several sheets of paper, a well-worn pencil, and, if you are like me, errors in your math! The proper model-building procedures (addition of variables, examination of residuals, transformations, etc.) are very likely to be completely ignored.

The in-house computer was very tight on time, partly because of "core-hogging" engineering problems. My initial request for help with routine statistical analysis met with serious questions in data processing. A programmer found a flowchart for a Monte Carlo simulation reliability analysis program in a handbook; it ran on for several pages. "Would I ever need anything like this?" Well, we were getting involved with a new guidance control. . . .

This led to the suggestion that all engineering programs be cleared from the in-house machine to time-sharing service. After some study, all agreed. The transition was made simpler by assistance from the computer time-sharing service company. In fact, they put the largest engineering programs on their machine before we even subscribed. As for statistics, the computer listing of statistics-related programs in the time-sharing library ran for well over 200 pages!

I was also exposed to training in computer programming with help from the time-sharing company. And, just as important, I became more familiar with the computer world's jargon along with the capabilities of computers. My ambitions grew as I learned not only of time-sharing possibilities but also of a more comprehensive assembly of programs. I researched the quality literature, magazines, case histories from other companies, ASQC literature on quality costs, and anything else I could find regarding computers in quality control. Gradually, the outline of a plan formed in my mind, a plan involving documenting the life history of parts from pilot run to obsolescence.

Before QC was really "ready," an opportunity arose to put part of my plan into effect. At about that time, phase I of an elaborate inventory control plan was entering the systems development stage.

The need for *some* advanced inventory control system was felt to be urgent. As our company grew from some $30 million in sales to nearly $60 million in a very short time, the obsolescence of the old system became glaringly obvious. One of the major duties of first-line supervision was to locate the correct parts for a job. This was increasingly difficult as volume increased and more and more parts were manufactured in-house. Added burdens were created by new product introductions until finally, with several thousand part numbers in existence, the old system collapsed completely. The first-generation replacement system was, like most forerunners of successful computer systems, an ingenious manual system. The basis of the system was the move tag, a simple document that was required before any material was moved or stored. Physical control of the parts within each storeroom was based on carefully partitioning the storeroom into definitive locations identified on each move tag.

Despite the vast improvement made by the new manual system, it was soon obvious that it was, at best, a stopgap measure with a number of limitations. One shortcoming was that 16 storerooms existed, not counting a myriad of work-in-process (WIP) locations, and a part might have locations in several storerooms. Another was the great difficulty involved in tying the manual storeroom system to computerized forecasting and ordering. In addition, there was the advantage to be gained from a computerized system in terms of minimizing inventory levels.

The plan for a *total* inventory control system was too much for our in-house mainframe computer system even without the big engineering programs. A decision had to be made as to which elements should be programmed first. For all the reasons given above, storeroom inventory control was selected. It is interesting that, at first, no provision was made to include QC in the planning. The main reason for the exclusion of QC was the feeling that most reject items were from WIP, an element due to be computerized in that never-never land known as "down the road. . . ." When confronted with the real facts (manually obtained), it was obvious that, in reality, big *dollars* were on "hold" status in stores—though goods in process represented high volume and many part numbers. I theorize that the mistaken belief resulted from the fact that our production control group was the prime mover, and rejects in process tend to be more dramatic disrupters of production schedules than rejects in stores. Our logic was accepted immediately and our participation in *all* future planning was assured and welcomed.

We approached the planning for on-line reject goods inventory control not only as an element in a larger system of defective material control but also as a part of a much more comprehensive program of *total* inventory control. The temptation to break off from the rest and push for our own freestanding system was great. In fact, we did explore such an option in detail but rejected it for several reasons:

1. True control involved not just a "paper" system but also physical control of reject goods. If, as an autonomous system implies,this meant removal of goods from the routine flow, it also meant that QA would be instantly in the storeroom business. This was felt to be undesirable.
2. Reject goods were often only delayed or diverted, not removed completely, from the routine flow. Thus, some goods might be repaired on the spot, and insistence on their physical segregation seemed counterproductive.
3. Possibly most important, we saw genuine opportunities for effective material control as a part of the larger system.

The enthusiasm exhibited by those participating in the system development was very impressive. Literally dozens of persons were involved over a period of 6 months, with the activities of all being coordinated by two employees assigned full-time coordination responsibility. One was from data processing and the other from production control. Yet, in spite of the sophistication of the spanking new "on-line real-time" computerized storeroom system, the foundation still rested on two factors already in place: committed, enthusiastic people and physical control of the material. The old manual move tag system was intended to supplement the computer-based system; however, there was one exception.

That exception was for rejected goods. The computer system (approximately 50 computer terminals strategically located for the 1600 employees) recorded for all parts just where the parts were, and this location identification included:

1. Which storerooms (of about 20) the parts were in. (Storerooms are identified by a four-digit number, and this was the sort used on the screen when one desired to see the storeroom inventory display by part.)
2. The location of the parts in the storerooms.
3. The identification numbers on the move tags.
4. How many of the parts were
   a. On the tag.
   b. In the location.
   c. In the storeroom.
   d. In stock (total of all storerooms' inventories).
5. When the last transaction was made.

It was item #3 that provided us with the opportunity to get better control of rejected goods. If any goods were rejected, the computer no longer referenced the move tag number. Instead, it referenced the *reject tag number*. As a futher identifier, all such tag numbers were preceded by the letter "R." This grouped them in the sort and separated them from the nonrejected storeroom supply. This made the control of reject items as good as the inventory control system combined with the quality department's control of reject tags. After a short learning period, the accuracy of the inventory control system was demonstrated and, due to an ongoing audit program by production control, we accepted the accuracy as a given. Control of reject tags, however, took a bit longer.

The problem with reject tag control had been growing for some time and had its roots in our paper-based control system. Each reject tag consisted of a cardboard copy attached to the material rejected, an original that was sent first to the "action department" (i.e., the department responsible for reworking or physically scrapping the product) and then to the quality control office, a carbon copy to be used by production and inventory control, and a third carbon for use by cost accounting in preparing the cost-of-quality report. Our problem was that no real incentive existed for the action department to return the original copy to quality control, which would close the feedback loop. Refer to Figure 5.4.

The new inventory control system made our paper system obsolete virtually overnight, giving us the opportunity to improve it. As mentioned above, the new system allowed identification of goods in any storeroom; thus production and inventory control no longer needed their copy of the reject tag. Likewise, cost accounting arranged to have a periodic listing of "negative transactions" (scrap), which made their copy obsolete. The question was what could be done with the two free copies of each reject tag. We (QC) immediately claimed one copy as our record of a reject tag outstanding. In a very short time we were shocked at the size of the outstanding reject tag pile. How much money was tied up in this inventory? I spoke with the vice-president of manufacturing and found that he, too, wondered. After discussions with other managers in manufacturing, it was agreed that this information, properly presented, could provide a useful scheduling tool. "Properly presented" meant a computer report sorted by action department, part number, and last operation completed. This would allow the work to be integrated with work in process. When we found that data processing was temporarily backlogged, we decided to produce the report on our time-sharing system. Cards were keypunched in house, but we had to look up part cost information and enter it by hand.

The reaction to the first report of rejects outstanding was nothing short of astounding. Meetings were held at high levels to discuss means of reducing the volume, responsibility for reduction sometimes being rather emotionally assigned to embarrassed managers. We in QC shared the embarrassment, as a large number of outstanding tags were found to be not outstanding at all, but filed away in the QC office. We had simply failed to remove the corresponding tag from the list of outstanding tags. About 40% of the rejects outstanding were cleared in the flurry of activity that followed the release of the first report. About half of this, 20%, was "paper" only as we cleared our files and supervisors scoured every desk drawer and file cabinet looking for "lost" reject tags.

With the unrelenting demands to reduce the volume of rejects outstanding, it was only natural that department supervisors wanted updated reports quickly to convince their superiors that results were indeed being achieved. But our report was scheduled to run monthly and

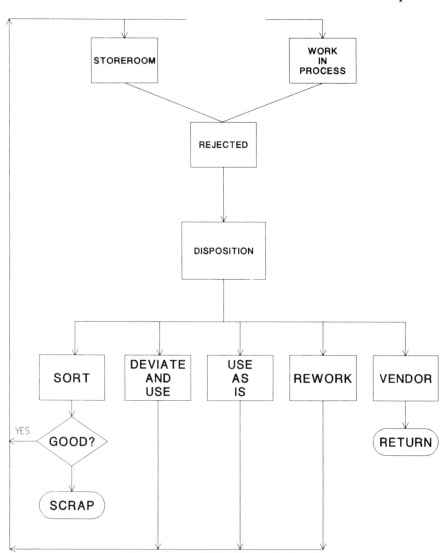

Figure 5.4   Reject material flow.

keypunch scheduling had to be revised. Complaints about lack of timeliness continued to mount until data processing rescheduled and agreed to run the program weekly in-house with data files maintained on line by QC department clerks. We went for several months with good control of rejects outstanding, and we have since returned to monthly reports with the concurrence of manufacturing.

The reports and systems discussed above were still a far cry from the "ultimate system" of internal failure costs, but they were enough of an improvement to move *external failure costs* to the forefront. We began a quiet campaign to convince our warranty group (who reported to the national service manager) that we could help them meet their objectives in reducing warranty costs if they could see their way clear to provide more detailed data on warranty failures. Up to that time the costs had been grouped in very broad categories—nearly the equivalent of "cars and trucks" in the auto industry. But our warranty, like that of most companies, was the accumulation of *part failures*, and we needed to know which parts were the big ones if real progress was to be made. It was finally agreed that a listing of the top ten warranty parts would be included with each monthly report. We would use three different measures: (1) highest dollars year-to-date, (2) highest quantity, and (3) most often claimed. As it turned out, the top ten was both too little and too much: too little because we still lacked data on other parts and still could not fill out our in-house part quality history with accurate field data, and too much because the top ten proved to be formidable problems indeed.

So much was involved in the solution of the top five problems that just the data storage and organization was more than we could handle! Our first response was to generate failure reports on the time-sharing service, but top problems meant frequent demands for the very latest reports and nearly continuous file maintenance. The burden was a real budget buster. Armed with the high cost of the time-sharing reports and the higher cost of field failures, we approached data processing for help. The response exceeded our wildest expectations; after hearing our stories (sometimes long and sad), they offered quality control 36 on-line data files! The files were to be maintained by QC personnel through the on-line network. We chose as our first two files the failure histories of the top two warranty problems. The transition from our manual system of edge-punched cards to the computer system was extremely easy and, shortly, we had fast, accurate data on our top field problems.

With the help of the new data, progress was made toward reducing the magnitude of the problems. Using our third file we put our outstanding reject report on line and made it possible to tie into the on-line storeroom and determine why parts were rejected, which department was responsible, where the parts were located, etc. File 4 was devoted to all parts that required receiving inspection. Any system user can determine whether a part requires inspection, the sampling plans, and the major characteristics to be inspected. File 4 also references any more detailed inspection procedures.

Our quality program also includes periodic part audits, which involve in-depth evaluation of a small sample. We have compiled the results and, over time, this has resulted in a valuable history. The audit history shows not only what parts are rejected but also which ones are accepted, a valuable measure of quality. It also enables us to produce data on the performance of specific machines and processes, measure the performance of the auditors, trace the quality of parts at various operations, and obtain insight into recurring problems.

SUMMARY

The chapter provided a definition of a QIS and its relationship to the broader subject of an MIS. Planning a QIS was described, along with the use of flowcharting as a planning aid. A real-life QIS implementation experience was presented as a case study. The importance of non-computerized elements was explained.

The chapter did not cover any specific hardware or software for QIS. Organizational concerns, such as distributed processing, were not covered. The relationship of QIS to material control systems like materials requirements planning (MRP) or just-in-time (JIT) was not discussed.

6

Probability and Statistics for Quality Control

The subjects of probability and statistics cover a tremendous amount of material. An individual could easily devote his or her life to either subject without ever exploring it entirely. This chapter will provide the reader with an overview of the elements of the subject critical to understanding basic quality control applications.

One of the most important problems facing anyone concerned with quality improvement is that all processes, no matter how well controlled, exhibit variation. The variation has as its source literally dozens of potential causes. Figure 6.1 shows a cause-and-effect diagram for a particular type of soldering defect, no solder.This particular diagram, based on an actual cause-and-effect diagram used in a statistical process control project, graphically displays only a few of the causes of this solder defect. Still, as you can see, there are dozens.

In a later chapter you will learn to construct cause-and-effect diagrams. For now, the important point is that any given "effect" is the result of a very large number of causes.

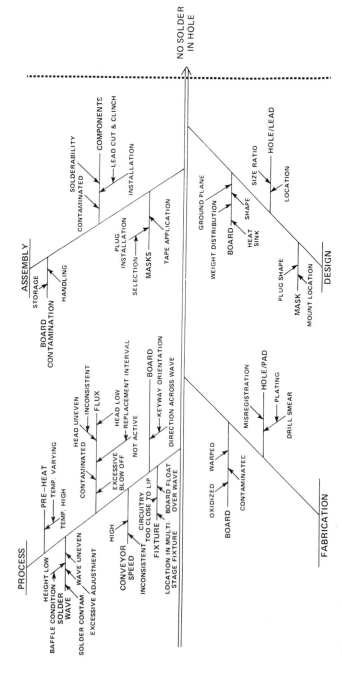

Figure 6.1 Cause of a solder defect.

Since we cannot realistically expect zero variation, and since there appear to be too many variables to keep track of, how are we to proceed in our improvement efforts? The answer lies in the use of statistical methods that allow us to understand variability and variation. With this understanding we can determine the appropriate course of action. This chapter describes the basic concepts of probability and statistics that are useful in quality control.

PROBABILITY

Probability theory forms the basis of statistical analysis. Probability is defined as follows:

Definition.   Probability is a set function P that assigns to each event A in sample space S a number $P(A)$ such that

1. $P(A)$ is equal to or greater than 0.
2. $P(S) = 1$.
3. If $A_1, A_2, \ldots$ are events that are mutually exclusive $[P(A_i \text{ and } A_j) = 0]$, then $P(A_1 \text{ or } A_2 \text{ or} \cdots) = P(A_1) + P(A_2) + \cdots$.

This rather academic definition can be made easier to understand with a simple example. Say I have a penny and I toss it in the air. There are two possible events: it will land with heads up or it will land with tails up. We will be practical here and ignore such extremely unlikely events as the coin landing on an edge or sailing off into space and never landing at all. The first point in the definition is that probability is a set function that assigns a number to each event. The number is called the probability of the event, and it must be at least zero. Futhermore, the sum of all the probabilities must be one. For our case, if the coin is fair, we have the same probability of seeing either a head or a tail. The last condition states that, if the events are mutually exclusive, the probability that one or the other will occur is the sum of the probabilities of the events occurring individually. The events were are considering are mutually exclusive, since we can't have both a head and a tail up at the same time. Thus $P(\text{head}) = .5$ and $P(\text{tail}) = .5$, and $.5 + .5 = 1$.

VENN DIAGRAMS

A simple graphical method that is useful for illustrating many probability relationships and other relationships involving sets is the Venn diagram. Figure 6.2 is Venn diagram of the example we have been discussing.

We say events A_1 and A_2 are *independent* if A_1 does not depend on A_2 and vice versa. When this is the case, the probability of two events occurring simultaneously is simply the product of the probabilities of the events occurring individually. For example, if 5% of our parts are defective, and if we purchase 50% of all our parts from vendor A and 50% of all the *defective* parts are from vendor A, then the probability of a part being both defective and from vendor A is

$$.05 \times .5 = .025$$

Figure 6.2 shows two circles that do not overlap, which means that they are mutually exclusive. If it is possible to get *both* events simultaneously, we have a different situation. Let's say for example that we have a group of 1000 parts (the sample space). Let event A_1 be that a part is defective and event A_2 be that the part is from vendor A. Since it is possible to find a defective part from vendor A, the two events can occur simultaneously. The Venn diagram in Figure 6.3 illustrates this.

If we have the situation illustrated by Figure 6.3, then we can no longer find the probability of either event occurring by just adding the individual probabilities. Instead, we must use the equation

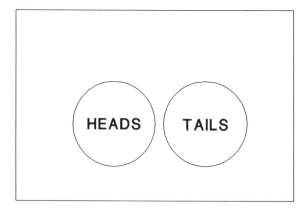

Figure 6.2 Venn diagram of two mutually exclusive events.

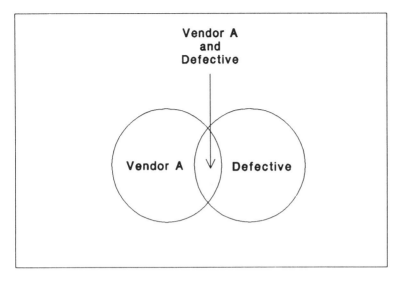

Figure 6.3 Venn diagram of two events that can occur simultaneously.

$$P(A_1 \text{ or } A_2) = P(A_1) + P(A_2) - P(A_1 \text{ and } A_2)$$

Returning to our previous example, let A_1 be the event that a part is defective and A_2 be the event that the part is from vendor A. Then, assuming the events are independent, the probability that the part is either defective or from vendor A is

$$.05 + .50 - (.05 \times .50) = .55 - .025 = .525$$

DISCRETE PROBABILITY DISTRIBUTIONS

There are two basic categories of probability distributions, discrete and continuous. A discrete variable is one that can take on only certain values; for example, a count can be only a nonnegative integer. A continuous variable can take on an infinite number of values, at least theoretically. For example, a temperature can be measured to as many decimal places as our measurement instrument can handle. Quality control work involves both discrete and continuous variables.

In applying probability theory to discrete variables in quality control, we frequently encounter the need for efficient methods of counting. One counting technique that is especially useful is combinations. The combination formula is

$$C_r^n = \frac{n!}{r!\,(n-r)!} \tag{1}$$

In the equation above, $n!$ (called "n factorial") means to multiply $n \times (n-1) \times (n-2) \times \cdots (n-n+1)$. Zero factorial is defined as 1 (i.e., $0! \equiv 1$). Equation one gives the number of unique ways in which you can arrange n objects taking them in groups of r objects at a time, where r is a nonnegative integer less than or equal to n. For example, to determine the number of combinations we can make with the letters X, Y, and Z in groups of two letters at a time, we note that $n = 3$ letters, $r = 2$ letters at a time, and use the above formula to find

$$C_2^3 = \frac{3!}{2!\,(3-2)!} = \frac{3 \times 2 \times 1}{(2 \times 1) \times (1)} = \frac{6}{2} = 3$$

The three combinations are XY, XZ, and YZ. Notice that this method does not count reversing the letters as separate combinations; that is, XY and YX are considered to be the same.

SAMPLING WITHOUT REPLACEMENT

Example. Let's look at a practical example of how combinations are applied to quality control. Assume we have received a lot of 12 parts from a distributor. We need the parts badly and are willing to accept the lot if it has fewer than three noncomforming parts. We decide to inspect only four parts since we can't spare the time to check every part. Checking the sample, we find one part that doesn't conform to our requirements. What is the probability that we would find one or fewer if the lot actually had three nonconforming parts?

The formula we need is known as the hypergeometric probability distribution. It is

$$P(x) = \frac{C_{n-x}^{N-m} C_x^m}{C_n^N} \quad \begin{array}{l} x \leqslant m \\ 0 \leqslant n \leqslant N \\ 0 \leqslant x \leqslant n \end{array} \quad (2)$$

In this equation N is the lot size, m is the number of defectives in the lot, n is the sample size, x is the number of defectives in the sample, and $P(x)$ is the probability of getting exactly x defectives in the sample. We must solve the above equation for $x = 0$ as well as $x = 1$, since we would also accept the lot if we had no defectives. The solution is shown below.

$$P(0) = \frac{C_{4-0}^{12-3} C_0^3}{C_4^{12}} = \frac{126 \times 1}{495} = .255$$

$$P(1) = \frac{C_{4-1}^{12-3} C_1^3}{C_4^{12}} = \frac{84 \times 3}{495} = \frac{252}{495} = .509$$

Adding the two probabilities tells us the probability that our "sampling plan" will accept lots of 12 with 3 nonconforming units. The plan of inspecting four parts and accepting the lot if we have 0 or 1 nonconforming has a probabiltiy of $.255 + .509 = .764$, or 76.4%, of accepting this "bad" quality lot. This is the "consumer's risk" for this sampling plan. Such a high sampling risk would be unacceptable to most people.

SAMPLING WITH REPLACEMENT

The approach discussed above is exact for evaluating sampling risks when each unit sampled is not returned to the lot before selecting the next sample unit. In practice, sample units are almost always selected at one time; therefore the above approach is correct for nearly every sampling situation involving lots or batches. However, it is not the correct method when sampling from a process in continuous production. In this case we are essentially sampling with replacement, since the "lot size" is the potential production, which is conceptually infinite. Assume that the process is producing some proportion of nonconforming units, which we will call p. If we are basing p on a sample, we find p by dividing the number of nonconforming units by the number of units sampled. The equation that will tell us the probability of getting x defectives in a sample of n units is

$$P(x) = C_x^n p^x (1 - p)^{n-x} \qquad 0 \leqslant x \leqslant n \tag{3}$$

Equation (3) is known as the *binomial probability distribution*. In addition to being useful as the exact distribution of nonconforming units for processes in continuous production, it is an excellent approximation to the hypergeometric probability distribution when the sample size is less than 10% of the lot size.

Example. A process is producing glass bottles on a continuous basis. Past history shows that 1% of the bottles have one or more flaws. If we draw a sample of 10 units from the process, what is the probability that there will be 0 nonconforming bottles?

Using the above information, $n = 10$, $p = .01$, and $x = 0$. Substituting these values into Equation (3) gives us

$$P(0) = C_0^{10} p^0 (1 - .01)^{10-0} = 1 \times 1 \times .99^{10}$$
$$= .904 = 90.4\%$$

Another way of interpreting the above result is that the sampling plan "inspect 10 units, accept the process if no nonconformances are found" has a 90.4% probability of accepting a process that is averaging 1% nonconforming units.

OCCURRENCES PER UNIT

Often in quality control we are not just concerned with *units* that don't conform to requirements; instead we are concerned with the number of nonconformances themselves. For example, say we are trying to control the quality of a computer. A complete audit of the finished computer would almost certainly reveal some nonconformances, even though they might be of minor importance (for example, a decal on the back panel might not be perfectly straight). If we tried to use the hypergeometric or binomial probability distribution to evaluate sampling plans for this situation, we would find they didn't work, because our lot or process would be composed of 100% nonconforming units. Obviously, we are interested not in the units per se, but in the nonconformances themselves.

The correct probability distribution for evaluating nonconformances is the *Poisson distribution*. The equation is

$$P(x) = \frac{u^x e^{-u}}{x!} \qquad x \geqslant 0 \tag{4}$$

In this equation u is the average number of nonconformances per unit, x is the number of nonconformances in the sample, and e is the constant 2.7182818; $P(x)$ gives the probability of exactly x occurrences in the sample.

Example. A production line is producing guided missiles. When each missile is completed, an audit is conducted by an Air Force representative and every nonconformance to requirements is noted. Even though any major nonconformance is cause for rejection, the prime contractor wants to control minor nonconformances as well. Such minor problems as blurred stencils and small burrs are recorded during the audit. Past history shows that on the average each missile has three minor nonconformances. What is the probability that the next missile will have 0 nonconformances?

We have $u = 3$, $x = 0$. Substituting these values into Equation (4) gives

$$P(0) = \frac{u^x e^{-u}}{x!} = \frac{3^0 (2.7182818)^{-3}}{0!} = \frac{1 \times .050}{1} = 0.05$$

In other words, 100% − 5% = 95% of the missiles will have at least one nonconformance.

The Poisson distribution, in addition to being the exact distribution for the number of nonconformances, is a good approximation to the binomial distribution in certain cases. To use the Poisson approximation, you simply let $u = np$ in Equation (4). Juran and Gryna (1980) recommend considering the Poisson approximation if the sample size is at least 16, the population size is at least 10 times the sample size, and the probability of occurrence p on each trail is less than 0.1. The major advantage of this approach is that it allows you to use the tables of the Poisson distribution, such as Table 2 in the Appendix.

CONTINUOUS PROBABILITY DISTRIBUTIONS

By far, the most common continuous distribution encountered in quality control work is the normal distribution. The probability density function for the normal distribution is given by the equation

$$y(x) = \frac{1}{\sigma\sqrt{2\pi}} e^{-(x-\mu)^2/2\sigma^2} \qquad -\infty < x < \infty \qquad (5)$$

In Equation (5), μ is the population average or mean and σ is the population standard deviation. The population mean is given by

$$\mu = \frac{1}{n} \sum_{i=1}^{n} x_i \qquad (6)$$

where the symbol Σ means "to sum." The mean is a measure of the central tendency of the data set. Let's say that we have the following values:

i	x_i
1	17
2	23
3	5

Then the mean is simply the sum of the three values divided by 3, the number of values, or

$$\mu = \frac{1}{3}(17 + 23 + 5) = \frac{45}{3} = 15$$

The standard deviation σ is computed as the square root of the variance σ^2 with the equations

$$\sigma^2 = \frac{1}{n} \sum_{i=1}^{n} (x_i - \mu)^2 \tag{7}$$

$$\sigma = \sqrt{\sigma^2} \tag{8}$$

The variance and standard deviation are both measures of dispersion or spread. Referring to the sample data above with a mean μ of 15, we compute σ^2 and σ as follows:

i	x_i	$x_i - \mu$	$(x_i - \mu)^2$
1	17	2	4
2	23	8	64
3	5	−10	100
		Sum	168

$$\sigma^2 = \frac{168}{3} = 56$$

$$\sigma = \sqrt{56} = 7.483$$

Usually we have only a sample and not the entire population. A *population* is the entire set of observations from which the *sample,* a subset, is drawn. In this case μ is computed in the same way as shown above, but the variance and standard deviation are computed using the formulas below.

$$s^2 = \frac{1}{n-1} \sum_{i=1}^{n} (x_i - \bar{X})^2 \tag{9}$$

$$s = \sqrt{s^2} \tag{10}$$

The differences between these two equations and those used for calculating the population variance and standard deviation are:

1. The Greek letter σ has been changed to *s* and μ has been changed to an *X* with a bar over it (pronounced *X* bar). Generally, Greek letters are used to represent population parameters and English letters are used to represent the sample statistics which estimate these parameters. The quantity *X* bar is computed in exactly the same way as μ.
2. The denominator of the variance equation has been changed from *n* to *n* − 1; *n* − 1 is known as the degrees of freedom. Using *n* − 1 makes the statistic *s* an unbiased estimator of the parameter σ. This simply means that the expected value of *s* = σ.

A complete discussion of statistical estimators is beyond the scope of this book. The subject is discussed in most introductory texts on mathematical statistics.

If we plotted a curve using Equation (5) for all values of *x* we would get a curve like the one shown in Figure 6.4. The normal distribution curve is sometimes called the bell curve because of its shape. This curve is said to be *symmetrical*, meaning that if a line is drawn at the mean of the distribution, the shape of the curve is the same on both sides of the line. It is also said to be *unimodal* because it has only one peak. The

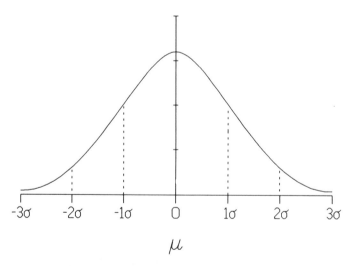

Figure 6.4 Normal distribution.

areas under the normal curve can be found by integrating Equation (5) (no closed-form solution exists to Equation 5), but more commonly tables are used. Table 1 in the Appendix contains areas under the normal curve. The table is indexed by using the Z transformation, which is

$$Z = \frac{x_i - \mu}{\sigma} \quad \text{or} \quad Z = \frac{x_i - \bar{x}}{s} \tag{11}$$

By using the Z transformation we convert any normal distribution into a normal distribution with the mean of 0 and a standard deviation of 1. Thus we can use a single normal table to find probabilities.

Example. The normal distribution is very useful in predicting long-term process yields. Assume we have checked the breaking strength of a gold wire bonding process used in microcircuit production and we have found that the process average strength is $9^{\#}$ and the standard deviation is $4^{\#}$. The process distribution is normal and the process is in statistical control. If the engineering specification is $3^{\#}$ minimum, what percentage of the process will be below the specification?

We must first compute Z using Equation (11):

$$Z = \frac{3 - 9}{4} = \frac{-6}{5} = -1.2$$

Entering Table 1 for $Z = -1.2$, we find that 11.51% of the area is below this Z value. Thus 11.51% of our breaking strengths will be below our low specification limit of $3^{\#}$. In quality control applications we usually try to have the average at least 3 standard deviations away from the specification; 4 standard deviations is preferred. To accomplish this we would have to improve the process by either raising the average breaking strength or reducing the process standard deviation, or both.

EXPONENTIAL DISTRIBUTION

Another distribution encountered often in quality control work is the exponential distribution. The exponential distribution is especially useful in analyzing reliability. The equation for the probability density function of the exponential distribution is

$$y = \frac{1}{\mu} e^{-x/\mu} \qquad x \geqslant 0 \tag{12}$$

Unlike the normal distribution, the exponential distribution has a highly skewed shape and there is a much greater area below the mean than above it. In fact, over 63% of the expontential distribution falls below the mean. Figure 6.5 shows an exponential distribution curve.

Unlike the normal distribution, the exponential distribution has a closed-form cumulative density function. The equation is shown in the example below.

Example. If a city water company averages 500 system leaks per year, what is the probability that the weekend crew, which works from 6 p.m. Friday to 6 a.m. Monday, will get no calls? Assume that the failure rate is in statistical control.

We have $\mu = 500$ leaks per year, which we must convert to leaks per hour. There are 365 days of 24 hours each in a year, or 8760 hours. Thus

Figure 6.5   Exponential distribution.

the mean time between failures (MTBF) is 8760/500 = 17.52 hours. There are 60 hours between 6 p.m. Friday and 6 a.m. Monday. Thus x_i = 60. Using Equation (13), we get

$$P(x \leqslant x_i) = \int_0^{x_i} \frac{1}{\mu} e^{-x/\mu} \, dx = 1 - e^{-x_i/\mu} \tag{13}$$

$$P(x \leqslant 60) = 1 - e^{-60/17.52} = 1 - e^{-3.425} = .967$$

This indicates that 96.7% of the weekends will receive some calls. Thus, the crew will get to loaf away 3.3% of the weekends.

STATISTICAL INFERENCE AND HYPOTHESIS TESTING

In almost all practical applications of statistics, including quality control applications, we are forced to make *inferences* about populations based on data from samples. In this chapter we have talked about sample averages and standard deviations; we have even used these numbers to make statements about future performance, such as long-term yields or potential failures. A problem arises that is of considerable practical importance: any estimate that is based on a sample has some amount of sampling error. This is true even if the sample estimates are the "best estimates" in the sense that they are *unbiased* estimators of the population parameters.

Statistical inference generally involves five steps:

1. Identifying the reference frame from which the sample is drawn and to which the inferences will apply
2. Formulating a hypothesis about the population or "state of nature"
3. Collecting a sample of observations from the population
4. Calculating statistics based on the sample
5. Either accepting or rejecting the hypothesis based on the predetermined acceptance criterion

There are two types of errors associated with statistical inference:

Type I error (α *error*): The probability that a hypothesis that is actually true will be rejected

Type II error (β *error*): The probability that a hypothesis that is actually false will be accepted

Type II errors are often plotted on what is known as an operating characteristics (OC) curve.

Another error that often occurs involves applying the conclusions to reference frames other than the frame from which the sample was drawn. This is especially common in quality control work concerning *future* reference frames, i.e., what will the process do tomorrow? It is important to remember that we cannot attach confidence levels to statements made about the future. Problems of this type are called analytical problems. Confidence intervals, hypothesis tests, and so on, can be applied only to enumerative problems where the reference frame is static. When making statements about the future performance of some process, the best we can do is describe our degree of belief in general terms (e.g., low, moderate, or high). The degree of belief depends on engineering and management knowledge combined with a demonstrated history of statistical control. Statistical methods such as confidence levels or hypotheses tests have no use because they are computed from a different reference frame.

So far in this chapter we have introduced a number of important statistics including the sample mean, the sample standard deviation, and the sample variance. These sample statistics are called *point estimators* because they are but single values used to represent population parameters. If the population is static and well defined, it is also possible to construct an interval about the statistics that has a predetermined probability of including the true population parameter. This interval is called a *confidence interval.* Interval estimation is an alternative to point estimation that gives us a better idea of the magnitude of the sampling error.

Confidence intervals are usually constructed as part of a *statistical test of hypotheses.* The hypothesis test is designed to help us make an inference about the true population value at a desired level of confidence.

As stated, confidence intervals and hypothesis tests apply to static and well-defined populations. However, the typical industrial situation is extremely dynamic. Thus, these methods have very little application in practical quality control work. Their main use is in acceptance sampling of isolated lots. Acceptance sampling is becoming obsolete as statistical process control (SPC) and parts-per-million quality levels become the norm.

NOTES

This brief introduction into the complex subject of statistics runs a very real risk of leaving you with enough knowledge to be dangerous. If you plan to apply statistical techniques to real-world situations, here are some words of advice:

1. Read the remainder of this book.
2. Take classes and seminars in statistics.
3. Use your common sense in interpreting results.

SUMMARY

This chapter defined probability and described some important probability concepts through the use of Venn diagrams. Sampling with and without replacement was evaluated and quality control examples provided using the hypergeometric and binomial probability distributions. The Poisson distribution and its use in quality control were described. Quality control applications of the normal distribution and exponential distribution were shown. Statistics that measure central tendency and dispersion were shown. A brief introduction to statistical inference and hypothesis testing in a quality control context was provided.

The chapter did not discuss the vast field of mathematical probability and statistics. No derivations or proofs were given for any of the formulas or theorems presented. While many methods of testing distributional assumptions exist, they were not presented. Finally, a wide variety of statistical techniques other than those presented in this chapter exist that can be and have been applied to quality control.

7

Statistical Process Control

Shewhart (1931) defined control as follows:
Definition of Control. A phenomenon will be said to be controlled when, through the use of past experience, we can predict, at least within limits, how the phenomenon may be expected to vary in the future. Here it is understood that prediction within limits means that we can state, at least approximately, the probability that the observed phenomenon will fall within the given limits.

The critical point in this definition is that control is *not* defined as the complete absence of variation. Control is simply a state where all variation is *predictable* variation. A controlled process isn't necessarily a sign of good management, nor is an out-of-control process necessarily producing nonconforming product. In all forms of prediction there is an element of uncertainty. We will call any unknown random cause of

---

Much of the material in this chapter was taken from *Statistical Process and Quality Control* by Thomas Pyzdek. Copyright 1986. Used by permission of the publisher, Quality Publishing, Inc., Tucson, Arizona.

variation of *chance cause* or a *common cause*; the terms are synonymous and will be used as such. If the influence of any particular chance cause is very small, and if the number of chance causes of variation is very large and relatively constant, we have a situation where the variation is predictable within limits. You can see from our definition above that a system such as this qualifies as a controlled system. Where Dr. Shewhart used the term "chance cause," Dr. W. Edwards Deming coined the term "common cause" to describe the same phenomenon. Both terms are encountered in practice.

Of course, not all phenomena arise from constant systems of common causes. At times the variation is caused by a source of variation that is not part of the constant system. These sources of variation were called *assignable causes* by Shewhart and *special causes* by Deming. Experience indicates that special causes of variation can usually be identified and corrected, leading to a process that is less variable.

Statistical tools are needed to help us identify the effects of special causes of variation. This leads to another definition:

Statistical Process Control. The use of statistical methods to identify the presence of special causes of variation in a process.

The basic rule of statistical process control is:

Variation from common cause systems should be left to chance, but special causes of variation should be identified and eliminated.

Figure 7.1 illustrates the need for statistical methods to determine the category of variation. The figure shows two time-ordered sequences of fractions defective on two different apparatus. Without statistical guidance it is likely that people would disagree on whether the observed variations were due to chance or whether some assignable cause of variation was affecting the process. Human beings have a very poor record of being able to discern the difference between chance variation and caused variation without such guidance.

The answer to the question "Should these variations be left to chance?" can only be obtained through the use of statistical methods. Figure 7.2 illustrates the basic concept.

In short, variation between the two "control limits" designated by the dashed lines in Figure 7.2 will be deemed as variation from the common cause system. Any variability beyond these fixed limits will be assumed to have come from special causes of variation. We will call any system exhibiting only common cause variation a statistically controlled system. It must be noted that the control limits are not simply

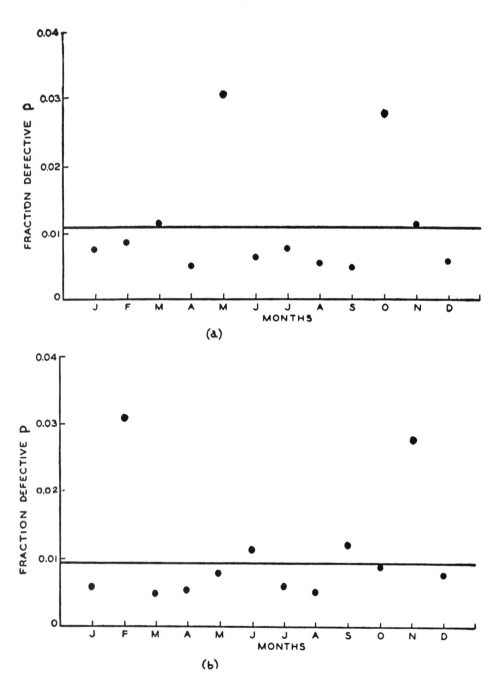

Figure 7.1 Should these variations be left to chance? (From Walter A. Shewhart, *Economic Control of Quality of Manufactured Product*, 1931, p. 13. Used by permission of the publisher, American Society for Quality Control).

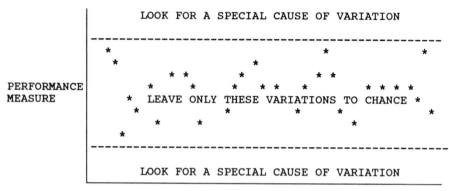

Figure 7.2 Types of Variation.

pulled out of the air; they are calculated from statistical theory. Figure 7.3 shows the control limits drawn on Figure 7.1. Notice that Figure 7.3a exhibits variations from special causes, while Figure 7.3b does not. This implies that the types of action needed to reduce the variability are of a different nature. In the case of Figure 7.3a variability reduction should begin with a search for assignable causes of variation, while the situation in Figure 7.3b indicates variability that is arising from chance causes of variation.

DISTRIBUTIONS

A central concept in statistical process control is that almost every measurable phenomenon is a statistical distribution. In other words, an observed set of data constitutes a sample of the effects of unknown common causes. It follows that, after we have done everything to eliminate special causes of variations, there will still remain a certain amount of variability exhibiting the state of control. Figure 7.4 illustrates the relationships between common causes, special causes, and distributions.

There are three basic properties of a distribution: location, spread, and shape. A distribution can be characterized by these three parameters. The location refers to the typical value of the distribution. In the previous chapter we discussed one measure of location, the mean. The spread of the distribution is the amount by which smaller values differ

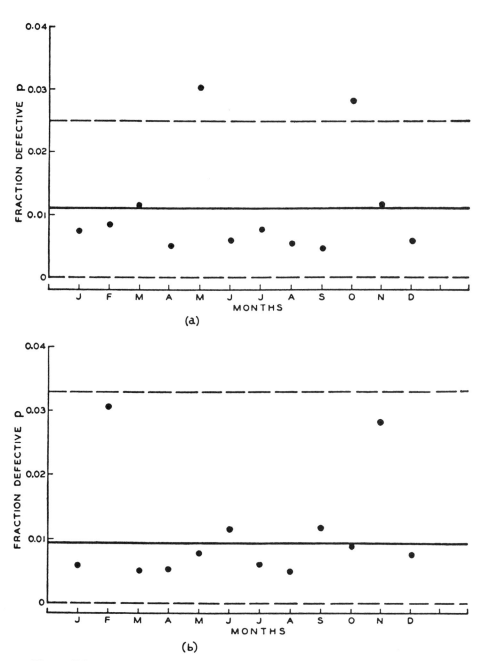

Figure 7.3 Charts from Figure 7.1 with control limits shown. (From Walter A. Shewhart, *Economic Control of Quality of Manufactured Product,* 1931, p. 16. Used by permission of the publisher, American Society for Quality Control.)

VARIATION: COMMON AND SPECIAL CAUSES

PIECES VARY FROM EACH OTHER:

BUT THEY FORM A PATTERN THAT, IF STABLE, IS CALLED A DISTRIBUTION:

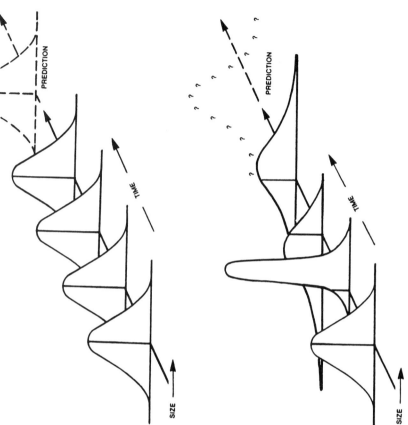

IF ONLY COMMON CAUSES OF VARI-
ATION ARE PRESENT, THE OUTPUT
OF A PROCESS FORMS A DISTRIBU-
TION THAT IS STABLE OVER TIME
AND IS PREDICTABLE:

IF SPECIAL CAUSES OF VARIATION
ARE PRESENT, THE PROCESS OUTPUT
IS NOT STABLE OVER TIME AND IS
NOT PREDICTABLE:

Figure 7.4 Distributions. (From *Continuing Process Control and Process Capability Improvement*, p. 4a. Copyright 1983. Used by permission of the publisher, Ford Motor Company, Dearborn, Michigan.)

from larger ones. The standard deviation and variance are measures of distribution spread. The shape of a distribution is its pattern—peakedness, symmetry, etc. A given phenomenon may have any one of a number of distribution shapes (the distribution may be bell-shaped, rectangular, etc.). In the previous chapter we discussed two common continuous distributions with distinctly different shapes, the normal distribution and the exponential distribution.

CENTRAL LIMIT THEOREM

The central limit theorem can be stated as follows:

> Irrespective of the shape of the distribution of the population or universe, the distribution of average values of samples drawn from that universe will tend toward a normal distribution as the sample size grows without bound.

It can also be shown that the average of sample averages equals the average of the universe and that the standard deviation of the averages equals the standard deviation of the universe divided by the square root of the sample size. Shewhart performed experiments that showed that small sample sizes (approximately four or five) were required to get approximately normal distributions from even wildly nonnormal universes. Figure 7.5 was created by Shewhart using samples of four units.

At this point you may be asking, so what? The practical implications of the central limit theorem are immense. Without the central limit theorem you would have to develop a separate statistical model for every nonnormal distribution encountered in practice. This would be the only way to determine whether the system was exhibiting chance variation. Because of the central limit theorem you can use *averages* of small samples to evaluate *any* continuous and many discrete process measurements using the normal distribution. The central limit theorem is the basis for the most powerful of statistical process control tools, Shewhart control charts.

REPRESENTING VARIATION IN A PROCESS

Measures of Central Tendency

As stated above, measurements from a process can be represented by a distribution. The distribution, in turn, can be characterized by its loca-

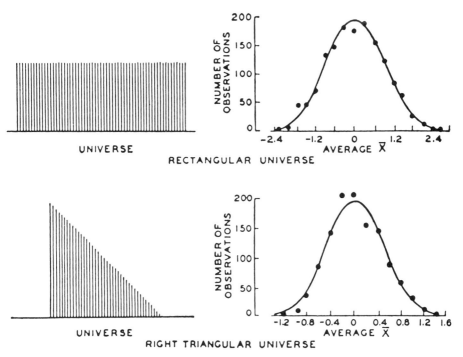

Figure 7.5 Illustration of the central limit theorem. (From Walter A. Shewhart, *Economic Control of Quality of Manufactured Product,* 1931. Used with permission of the publisher, American Society for Quality Control.)

tion, spread, and shape. The location of a distribution was described as its "typical value." When we talk about the *location* of a distribution in this book we are referring to the central tendency of the distribution. Remember that we said a distribution is the pattern that remains after we have eliminated all special causes of variation. This assumes that we have obtained a reasonably large number of samples, analyzed their pattern, and found that they exhibit variation that could arise from a common cause system. What might a typical value from this common cause system be? Intuitively, the value that divides the data into two equal parts is a typical value. This is known as the median. The median is calculated as follows.

Steps for finding the median

1. Sort the data from smallest to largest, or largest to smallest.
2. If the count of values is an odd number, then the median is the mid-

dle value. For example, if you have five numbers the median is the third value. The example below illustrates this.

Data: 4.3, 3.2, 1.1, 4.0, 2.4

Rank	Value
1	1.1
2	2.4
3	3.2
4	4.0
5	4.1

← 3rd value = median = 3.2

3. If the count of values is an even number, then the median is obtained by dividing the two middle values by 2. For example,

Data: 4.3, 1.1, 4.0, 2.4

Rank	Value
1	1.1
2	2.4
3	4.0
4	4.1

these are the middle values

$$\text{Median} = \tilde{X} = \frac{2.4 + 4.0}{2} = \frac{6.4}{2} = 3.2 \tag{17}$$

Note the symbol for the median, "x tilde."

Another typical value might be the value that occurs most often. The statistical term for this value is the *mode*. Figure 7.6 illustates the mode.

The final typical value we will consider here is the *mean*, or average value. The mean is the center of gravity of a distribution and it is found by adding up all the values and dividing the sum by the count. Mathematically, here is how we write this:

$$\text{Mean} = \bar{X} = \frac{1}{n} \sum_{i=1}^{n} x_i = \frac{\text{sum of values}}{\text{number of values}} \tag{18}$$

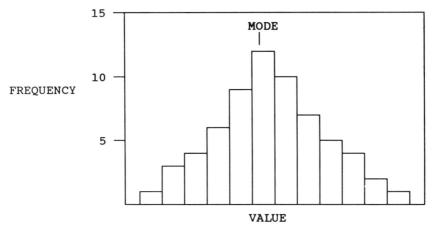

Figure 7.6 The Mode.

The mean is by far the most commonly used typical value, or measure of central tendency of a distribution. You may wish to compare Equation (18), the sample mean, with Equation (6), the population mean.

Measures of Dispersion

The second property of a distribution that we will want to measure is its spread, or dispersion. There are a number of useful statistics that quantify a distribution's spread. The variance and standard deviation, described in the previous chapter, have a number of desireable mathematical properties. However, the easiest measure of dispersion to calculate is the range.

$$\text{Range} = R = \text{largest value} - \text{smallest value} \tag{19}$$

Fortunately, when the distribution is from a constant system of chance causes, that is, the distribution is from a process in a state of statistical control, we can use the average range to obtain a very satisfactory estimate of σ. The relationship is given by

$$\hat{\sigma} = \frac{\bar{R}}{d_2} \tag{20}$$

where d_2 is a constant. Table 6 gives d_2 values for various sample sizes. Please note that Equation (20) holds *if and only if the process is in statistical control.*

PREVENTION VERSUS DETECTION

A process control system is essentially a feedback system. Four main elements are involved: the process itself, information about the process, action taken on the process, and action taken on the output from the process. The way these elements fit together is shown by Figure 7.7.

By the process, we mean the whole combination of people, equipment, input materials, methods, and environment that work together to produce output. The performance information is obtained, in part, from evaluation of the process output. The output of a process includes more than product; it also includes information about the operating state of the process such as temperature and cycle times. Action taken on a *process* is future-oriented in the sense that it will affect output yet to come. Action on the *output* is past-oriented because it involves detecting out-of-specification output that has already been produced.

There has been a tendency in the past to concentrate attention on the past-oriented strategy of inspection. With this approach we wait until output has been produced; then the output is inspected and either

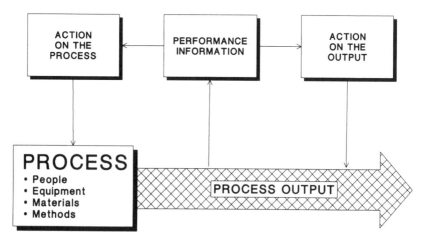

Figure 7.7 A process control system.

accepted or rejected. It should be obvious that this does nothing to prevent substandard output in the future. Statisical process control takes us in a completely different direction: improvement in the future. A key concept is that *the smaller the variation around the target, the better.* Thus, according to this school of thought, it is not enough to merely meet the requirements. Continuous improvement is called for. The concept of never-ending, continuous improvement is at the heart of SPC.

CAUSE-AND-EFFECT DIAGRAMS

Process improvement involves taking action on the causes of variation. With most practical applications the number of possible causes for any problem can be huge. Dr. Kaoru Ishikawa developed a simple method of graphically displaying the causes of any given quality problem. His method is called by several names: the Ishikawa diagram, the fishbone diagram, or the cause-and-effect diagram.

Cause-and-effect diagrams are tools that are used to organize and graphically display all of the knowledge a group has related to a particular problem. Usually, the steps are

1. Develop a flowchart of the area to be improved.
2. Define the problem to be solved.
3. Brainstorm to find all possible causes of the problem.
4. Organize the brainstorming results in rational categories.
5. Construct a cause-and-effect diagram that accurately displays the relationships of all the data in each category.

Once these steps are complete, constructing the cause-and-effect diagram is very simple. The steps are

1. Draw a box on the far right-hand side of a large sheet of paper and draw a horizontal arrow that points to the box. Inside the box, write the description of the problem you are trying to solve.
2. Write the names of the categories above and below the horizontal line. Think of these as branches from the main trunk of the tree.
3. Draw in the detailed cause data for each category. Think of these as limbs and twigs on the branches.

A good cause-and-effect diagram will have many twigs. (See Figure 7.8). If your cause-and-effect diagram doesn't have a lot of smaller branches and twigs, it shows that the understanding of the problem is superficial. Chances are you need the help of someone outside your

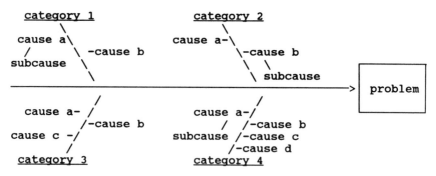

Figure 7.8 Cause-and-effect diagram.

group to aid in the understanding, perhaps someone more closely associated with the problem.

Cause-and-effect diagrams come in several basic types. The *dispersion analysis type* is created by repeatedly asking "Why does this dispersion occur?" For example, we might want to know why all of our fresh peaches don't have the same color.

The *production process class* cause-and-effect diagram uses production processes as the main categories, or branches of the diagram. In Figure 7.9 processes are shown joined by the horizontal line.

The *cause enumeration* cause-and-effect diagram, as shown in Figure 7.8, simply displays all possible causes of a given problem grouped according to rational categories. This type of cause-and-effect diagram lends itself readily to the brainstorming approach.

Cause-and-effect diagrams have a number of uses. Creating the diagram is an education in itself. Organizing the knowledge of the

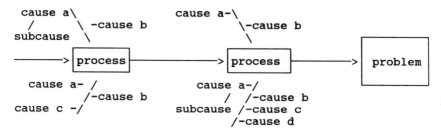

Figure 7.9 Production process class cause-and-effect diagram.

group serves as a guide for discussion and frequently inspires more ideas. The cause-and-effect diagram, once created, acts as a record of your research. Simply record your tests and results as you proceed. If the true cause is found to be something that wasn't on the original diagram, write it in. Finally, the cause-and-effect diagram is a display of your current level of understanding. It shows the existing level of technology as understood by the team. It is a good idea to post the cause-and-effect diagram in a prominent location for all to see.

CEDAC

A variation of the basic cause-and-effect diagram, developed by Dr. Ryuji Fukuda of Japan, is the cause-and-effect diagrams with the addition of cards, or CEDAC. The main difference is that the group gathers ideas outside the meeting room on small cards, as well as in group meetings. The cards also serve as a vehicle for gathering input from people who are not in the group; they can be distributed to anyone involved with the process. Often the cards provide more information than the brief entries on a standard cause-and-effect diagram. The cause-and-effect diagram is built by actually placing the cards on the branches. The reader is referred to Fukuda (1983) for additional information on CEDAC.

DESCRIPTIVE DATA ANALYSIS

Statistical process control, technically, uses only inferential statistical techniques. However, in practice some descriptive techniques have proved very useful in SPC efforts. A descriptive statistical technique merely reorganizes the data without making any inference regarding population parameters. Descriptive techniques are explorative in nature. Two descriptive techniques are described here: Pareto analysis and histograms. The information is presented in a "procedure format" to make it easier to see how the technique is actually used. This format will be used for all the SPC techniques presented in this book.

Pareto Analysis

Definition.   Pareto analysis is the process of ranking opportunities to determine which of many potential opportunities should be pursued first. It is also known as "separating the vital few from the trivial many."

Usage. Pareto analysis should be used at various stages in a quality improvement program to determine which step to take next. It is used to answer such questions as "What department should have the next SPC team?" or "On what type of defect should we concentrate our efforts?"

How to Perform a Pareto Analysis

1. Determine the classifications (Pareto categories) for the graph. If the desired information does not exist, obtain it by designing check-sheets and logsheets.
2. Select a time interval for analysis. The interval should be long enough to be representative of typical performance.
3. Determine the total occurrences (i.e., cost, defect counts, etc.) for each category. Also determine the grand total. If several categories account for only a small part of the total, group these into a category called "other."
4. Compute the percentage for each category by dividing the category total by the grand total and multiplying by 100.
5. Rank order the categories from the largest total occurrences to the smallest.
6. Compute the "cumulative percentage" by adding the percentage for each category to that of any preceding categories.
7. Construct a chart with the left vertical axis scaled from zero to at least the grand total. Put an appropriate label on the axis. Scale the right vertical axis from 0 to 100%, with 100% on the right side being the same height as the grand total on the left side.
8. Label the horizontal axis with the category names. The left-most category should be the largest, next the second largest, and so on.
9. Draw in bars representing the amount of each category. The height of the bar is determined by the left vertical axis.
10. Draw on line that shows the cumulative percentage column of the Pareto analysis table. The cumulative percentage line is determined by the right vertical axis.

Example of Pareto Analysis. The following data have been recorded for peaches arriving at Super Duper Market during August.

Problem	Peaches lost
Bruised	100
Undersized	87
Rotten	235
Underripe	9
Wrong variety	7
Wormy	3

The Pareto table and diagram for the above data are presented below.

Rank	Category	Count	Percentage	Cumulative %
1	Rotten	235	53.29	53.29
2	Bruised	100	22.68	75.97
3	Undersized	87	19.73	95.70
4	Other	19	4.31	100.01

Note that, as often happens, the final percentage is slightly different from 100%. This is due to round-off error and is nothing to worry about.

Histograms

Definition. A plot of a frequency distribution in the form of rectangles whose bases are equal to the cell interval and whose areas are proportional to the frequencies. In simpler terms, a histogram is a pictorial representation of a set of data.

Usage. Histograms are used to determine the shape of a data set. Also a histogram displays the numbers in a way that makes it easy to see the dispersion and central tendency and to compare the distribution to requirements. Histograms can be valuable troubleshooting aids. Comparisons between histograms from different machines, operators, vendors, etc. often reveal important differences.

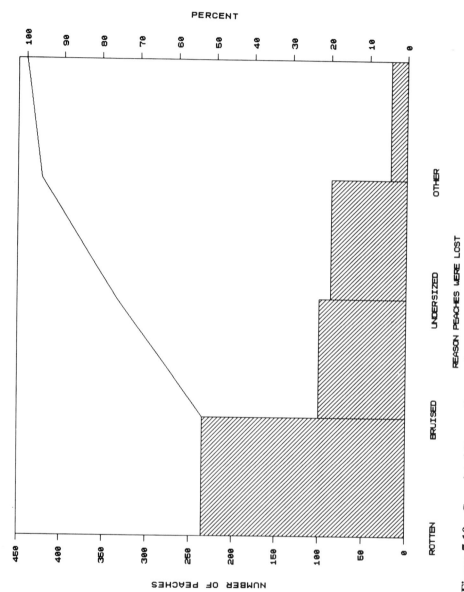

Figure 7.10 Completed Pareto diagram for lost peaches.

How to Construct a Histogram

1. Find the largest and the smallest value in the data.
2. Compute the range (we use the letter R to represent the range throughout this book) by subtracting the smallest value from the largest value. [See Equation (19).]

R = largest value − smallest value

3. Select a number of cells for the histogram. The table below provides some guidelines:

Number of values	Number of cells
100 or less	7–10
101–200	11–15
201 or more	13–20

The final histogram may not have exactly the number of cells you choose here, because of rounding.

4. Determine the width (W) of each cell. The cell width W is computed from the formula

$$W = \frac{R}{\text{number of cells}} \tag{21}$$

where R is from Equation (19).

5. The number W is a starting point. You should round W to a convenient number.

6. Compute "cell boundaries." A cell is a range of values, and cell boundaries define the start and end of each cell. To avoid ambiguity, cell boundaries should have one more decimal place than the raw data values in the data set. The low boundary of the first cell must be less than the smallest vaue in the data set. Other cell boundaries are found by adding W to the previous boundary. Continue until the upper boundary of a cell is larger than the largest value in the data set.

7. Compute the cell midpoint for the first cell by adding the lower and upper boundaries of the first cell and dividing the total by 2. The midpoint of subsequent cells can be found by adding W to the previous midpoint.

8. Now go through the raw data and determine the cell into which each value falls. Mark a tick in the appropriate cell.

9. Count the ticks in each cell and record the count, also called the frequency, to the right of the tick marks.

10. Construct a graph from the table. The vertical axis of the graph will show the frequency in each cell. The horizontal axis will show the cell boundaries.

11. Draw bars representing the cell frequencies. The bars should all be the same width, and the height of the bars should equal the frequency in the cell.

Histogram Example.   Assume you have the following data on the size of a metal rod.

Row	Data				
1	1.002	0.995	1.000	1.002	1.005
2	1.000	0.997	1.007	0.992	0.995
3	0.997	1.013	1.001	0.985	1.002
4	0.990	1.008	1.005	0.994	1.012
5	0.992	1.012	1.005	0.985	1.006
6	1.000	1.002	1.006	1.007	0.993
7	0.984	0.994	0.998	1.006	1.002
8	0.987	0.994	1.002	0.997	1.008
9	0.992	0.988	1.015	0.987	1.006
10	0.994	0.990	0.991	1.002	0.988
11	1.007	1.008	0.990	1.001	0.999
12	0.995	0.989	0.982*	0.995	1.002
13	0.987	1.004	0.992	1.002	0.992
14	0.991	1.001	0.996	0.997	0.984
15	1.004	0.993	1.003	0.992	1.010
16	1.004	1.010	0.984	0.997	1.008
17	0.990	1.021*	0.995	0.987	0.989
18	1.003	0.992	0.992	0.990	1.014
19	1.000	0.985	1.019	1.002	0.986
20	0.996	0.984	1.005	1.016	1.012

The frequency table for the above data is shown below and the histogram is given in Figure 7.11.

Cell	Cell boundaries	Midpoint	Tally	Frequency
1	0.9815–0.9855	0.9835	\\\\\\\\	8
2	0.9855–0.9895	0.9875	\\\\\\\\\	9
3	0.9895–0.9935	0.9915	\\\\\\\\\\\\\\\\\	17
4	0.9935–0.9975	0.9955	\\\\\\\\\\\\\\\\	16
5	0.9975–1.0015	0.9995	\\\\\\\\\	9
6	1.0015–1.0055	1.0035	\\\\\\\\\\\\\\\\\\\	19
7	1.0055–1.0095	1.0075	\\\\\\\\\\\	11
8	1.0095–1.0135	1.0115	\\\\\\	6
9	1.0135–1.0175	1.0155	\\\	3
10	1.0175–1.0215	1.0195	\\	2

Average and Range Charts

Definition. Average charts are statistical tools used to evaluate the central tendency of a process over time. Range charts are statistical

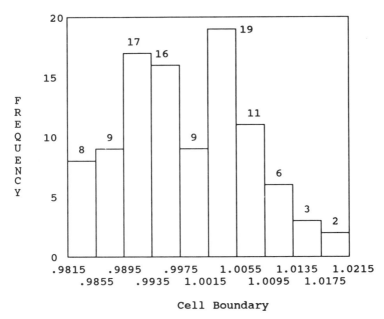

Figure 7.11 Histogram.

tools used to evaluate the dispersion or spread of a process over time.

Usage. Average charts answer the question, "Has a special cause of variation caused the central tendency of this process to change over the time period observed?" Range charts answer the question, "Has a special cause of variation caused the process distribution to become more or less erratic?" Range charts are nearly always used in conjunction with another SPC chart that evaluates the central tendency of the process. Average and range charts can be applied to any continuous variable such as weight or size.

How to Prepare and Analyze Average and Range Charts

1. Determine the subgroup size and sampling frequency. Typically, subgroups of four to six units are sufficient and subgroups of five are most common. The subgroup size affects the sensitivity of the control chart. Smaller subgroups give a control chart that is less sensitive to changes in the process, while larger subgroups are too sensitive to small, unimportant changes. Another factor to consider is the central limit theorem, for distributions that are extremely nonnormal in shape subgroups of at least four are necessary to assure that averages will be approximately normally distributed. Selection of subgroups should be planned to minimize possible variation; this is usually accomplished by selecting consecutive units. Sampling should be frequent enough to detect the effect of special causes while the special cause itself can still be identified. This varies a great deal from process to process, but as a rule you should average about 1 out-of-control point per typical control chart of 25 groups. If you have more than that, increase the sampling frequency. If you have fewer, reduce the sampling frequency. See Montgomery (1986) on using cost models to establish sampling frequency and subgroup size.

2. Collect data from 20 to 25 subgroups; at least 100 individual values are recommended. While the data are being collected, minimize disturbances to the process. If a process change is unavoidable, develop a system for recording changes so that their effect can be determined.

3. Perform the necessary calculations. The flowchart in Figure 7.12 is reprinted from the June 1986 ASQC Statistics Division Newsletter. The form was developed by Barry Griffin of Valparaiso University. Use of the form is quite simple. Each arithmetic instruction in the com-

CONTROL CHART CALCULATIONS ($\bar{X}$ AND R CHARTS)

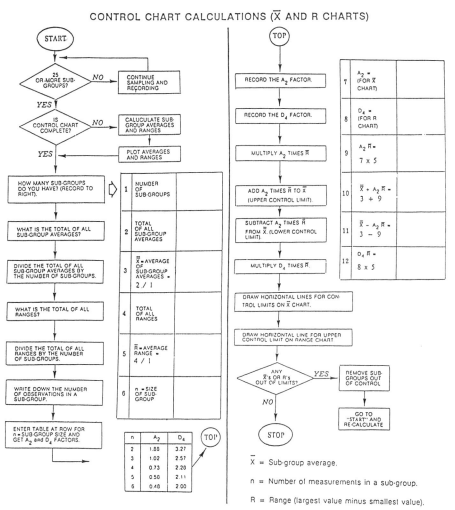

Figure 7.12 Flowchart for X-bar and R chart calculations.

putational sequence is accompanied by a rectangle divided into three segments: a boldface identification number, a description of the data element or instruction, and a blank space for recording the resulting data. Boldface numbers written beneath the description or instruction in the second segment refer you back to the corresponding rectangles. In the descriptive portion of rectangle number 5, for example, **4/1** means that you should divide the data element written in rectangle number 4 by the data element in rectangle 1. By stepping through this simple logic, you will calculate control limits for both X-bar and R charts by the time you complete rectangle number 12.

4. Interpretation of control charts. The interpretation of control charts for the average and the range involves two considerations: freaks and nonrandom patterns. Either case represents the presence of a special cause of variation. Freaks are detected by comparing each individual subgroup average and range to the control limits computed in rectangles 10, 11, and 12 in the flowchart. The upper control limit for the averages is the value in rectangle 10; the formula is

$$\text{UCL}_{\bar{x}} = \bar{\bar{X}} + A_2\bar{R} \qquad (22)$$

The A_2 value is a constant from Table 6 of the Appendix. All control limits are 3σ limits; however, we are using the average range instead of σ as a measure of spread. The A_2 constants account for this, and they also account for the fact that we are plotting averages instead of individual values. Thus, the upper control limit is in reality 3 standard deviations of the mean above the grand average. The situation is analogous for the lower control limit for the averages, which is the value in rectangle 11. The equation for the lower control limit is

$$\text{LCL}_{\bar{x}} = \bar{\bar{X}} - A_2\bar{R} \qquad (23)$$

The upper control limit for the ranges is given in rectangle 12; for subgroups of less than 7 the lower control limit for the ranges is 0. The control limits for the range chart are

$$\text{LCL}_R = D_3\bar{R} \qquad (24)$$
$$\text{UCL}_R = D_4\bar{R} \qquad (25)$$

As before, the D_3 and D_4 values are in Table 6 and are designed so that the control limits are 3σ on either side of the average. In this case we are talking about 3σ of the range. Note that for subgroups smaller than 7, D_3 = 0, which gives a lower control limit of zero. Because of this the table included with the flowchart, which only goes up to subgroup size 5, doesn't give D_3 values.

5. Control charts and the normal distribution. Under the assumption that the distribution of the statistic used to control the process distribution is normal, the probability of a single point exceeding a control limit, if the process is in statistical control, is 2 in 741. This is illustrated in Figure 7.13. However, the basis of control limits is a practical matter as well as a statistical one. As stated by Shewhart (1931), "Hence the method for establishing allowable limits of variability in a statistic Ω depends upon theory to furnish the expected value Ω-bar and the standard deviation σ_Ω of the statistic Ω and upon empirical evidence to justify the choice of limits Ω-bar $\pm t\sigma_\Omega$." In practical use, control charts are not a statistical test of hypothesis. Control charts provide an operational definition of a special cause of variation. The diagram shows that the average or range chart can be thought of as a normal distribution "stretched out in time." With this in mind we can also analyze the *patterns* on the control charts to see if they might be attributed to a special cause of variation. To do this we begin with Figure 7.13 and add two more lines on each side of the grand average, at 1 sigma and 2 sigma, respectively. Since, as shown on the diagram, the control limits are at plus and minus 3 sigma, finding the 1 and 2 sigma lines is as simple as

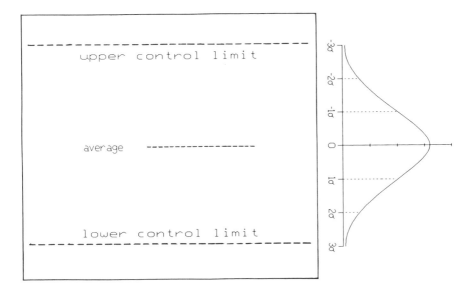

Figure 7.13 How a control chart is related to the normal distribution.

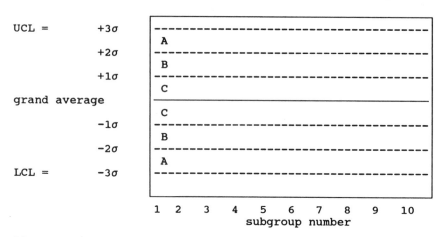

Figure 7.14 Zones on a control chart.

dividing the zone between the grand average and either control limit into thirds. This divides each half of the control chart into three zones. The three zones are labeled A, B, and C in Figure 7.14. Using the guide in Figure 7.14, tests for nonrandom patterns are applied. Remember, the existence of a nonrandom pattern means that a special cause of variation was (or is) probably present. Our X-bar control chart tests are shown in Figure 7.15 (Nelson, 1984). Note that when a point responds

Row			Data		
1	1.002	0.995	1.000	1.002	1.005
2	1.000	0.997	1.007	0.992	0.995
3	0.997	1.013	1.001	0.985	1.002
4	0.990	1.008	1.005	0.994	1.012
5	0.992	1.012	1.005	0.985	1.006
6	1.000	1.002	1.006	1.007	0.993
7	0.984	0.994	0.998	1.006	1.002
8	0.987	0.994	1.002	0.997	1.008
9	0.992	0.988	1.015	0.987	1.006
10	0.994	0.990	0.991	1.002	0.988
11	1.007	1.008	0.990	1.001	0.999
12	0.995	0.989	0.982	0.995	1.002
13	0.987	1.004	0.992	1.002	0.992
14	0.991	1.001	0.996	0.997	0.984
15	1.004	0.993	1.003	0.992	1.010

Figure 7.15 Tests for out-of-control patterns on control charts. (From Tests for Out of Control Patterns on Control Charts. Copyright 1984, American Society for Quality Control, Inc., Milwaukee, Wisconsin. Reprinted by permission from American Society for Quality Control.)

Row			Data		
16	1.004	1.010	0.984	0.997	1.008
17	0.990	1.021	0.995	0.987	0.989
18	1.003	0.992	0.992	0.990	1.014
19	1.000	0.985	1.019	1.002	0.986
20	0.996	0.984	1.005	1.016	1.012

to an out-of-control test it is marked with an "x" to make the interpretation of the chart easier. Using this convention, the patterns on the control charts can be used as a aid in troubleshooting.

Using the data, we complete the form as shown in Figure 7.16. Note that when completing the form for rectangles 10 and 11 we carried two more decimal places than the raw data contain. It is recommended that the average be rounded to one decimal place more than the raw data. The range control limit in rectangle 12 has one more decimal place than the raw data. The above conventions eliminate the ambiguity that might result if a sample average or range fell exactly on the control limit. The completed control charts are shown in Figure 7.17. There are some minor differences between the control chart limits and the figure 17.16 worksheet. These are due to roundoff errors.

Control Charts for Proportion Defective (*p* Charts)

Definition. *p* charts are statistical tools used to evaluate whether the proportion nonconforming produced by a process demonstrates a state of statistical control. The determinations are made through comparison of the values of some statistical measures for an ordered series of samples, or subgroups, with control limits.

Usage. *p* charts can be applied to any variable where the appropriate performance measure is a unit count. These charts answer the question, "Has a special cause of variation caused this process to produce an abnormally large or small number of defective units over the time period observed?"

How to Prepare and Analyze p Charts

1. Determine the sampling frequency. Sampling should be frequent enough to detect the effect of special causes while the special cause itself can still be identified. This varies a great deal from process to process, but as a rule you should average about 1 out-of-control point per typical control chart of 25 groups. If you have more than that, in-

CONTROL CHART CALCULATIONS ($\overline{X}$ AND R CHARTS)

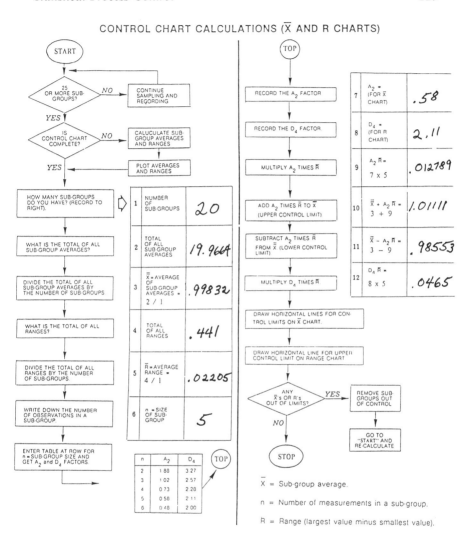

Figure 7.16 Example of completed Xbar and R chart work sheet.

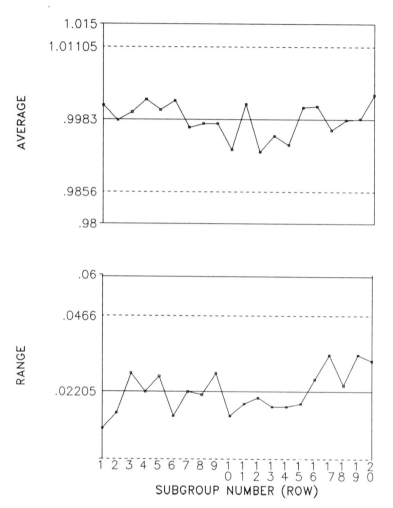

Figure 7.17   Example of completed *X*-bar and *R* Charts.

crease the sampling frequency. If you have fewer, reduce the sampling frequency.

2. Collect data from 25 to 30 subgroups. While the data are being collected, minimize disturbances to the process. If a process change is unavoidable, develop a system for recording changes so that their effect can be determined.

3. Establish the control limits. Perform the necessary calculations. The worksheet (Figure 7.18) will be helpful in calculating the control limits. Note that the control limits vary as the sample size varies. Thus, for p charts there are separate control limits for every subgroup. Also note that control limits should be based on data taken when the process was in control. This means that you should identify the special causes of out-of-control subgroups and drop these groups from your calculations. The worksheet solves three equations: the average proportion defective, the upper control limit for the fraction defective, and the lower control limit for the fraction defective. The upper and lower control limits must be solved for each subgroup since the standard deviation of a proportion varies as the sample size varies. The equations being solved by the worksheet are

$$\bar{p} = \frac{\text{total number defective}}{\text{total number inspected}} \tag{26}$$

$$\text{LCL}_P = \bar{p} - 3\sqrt{\frac{\bar{p}(1 - \bar{p})}{n_i}} \tag{27}$$

$$\text{UCL}_P = \bar{p} + 3\sqrt{\frac{\bar{p}(1 - \bar{p})}{n_i}} \tag{28}$$

where n_i is the size of each individual subgroup. As with average and range charts, the control limits are set symmetrically 3σ on either side of the average. However, with p charts the lower control limit sometimes falls below zero; in these cases it is simply set equal to zero. If the upper control limit is greater than one it is set equal to one.

4. Analysis of p charts. As with all control charts, a special cause is probably present if there are any points beyond either the upper or the lower control limit. Analysis of p chart patterns between the control limits is extremely complicated because the sample size varies and the distribution of the number defective varies with the sample size. Because of this, pattern analysis as described for X-bar charts is not done for p charts. A method known as a stabilized p chart converts all of the proportions to standard deviation units and *does* allow constant control limits and pattern analysis. The reader is referred to Duncan (1974) for additional information on stabilized p charts.

1. Collect information on 20 or more subgroups.
2. Record the size of each subgroup in column 2 of the worksheet on the next page.
3. Record the number defective in each subgroup in column 3 of the worksheet.
4. Compute the subgroup proportion defective, p, by dividing the subgroup number defective (column 3) by the subgroup size (column 2). Record the result in column 3. NOTE: When using the worksheet, any number in parentheses ( ) refers to a column number, e.g., $(3) \div (2)$ means to divide the value in column 3 by the value in column 2.
5. Compute the square root of each subgroup size and record the result in column 5. NOTE: The notation sqrt{x} means to take the square root of x, the value in the braces { }.
6. Add the sample sizes for all of the subgroups. Record the result at the bottom of column 2 and in the box below labeled "n = sum of all subgroups."
7. Add the sample defectives for all of the subgroups. Record the result at the bottom of column 2 and in the box below labeled "d = sum of all defectives."
8. Complete the calculations for all of the remaining boxes on this page. The numbers that go in each box are always computed in a previous step.

```
n = sum of all subgroups      [        ]

d = sum of all defectives     [        ]

pbar = d÷n = [        ]  ÷  [        ]  =  [        ]

qbar = 1 - pbar  = 1  -  [        ]  =  [        ]

A = pbar x qbar  =  [        ]  x  [        ]  =  [        ]

B = sqrt ( A )  = sqrt ( [        ] )= [        ]

C = 3 x B  = 3 x  [        ]  =  [        ]
```

Figure 7.18 *p* Chart Worksheet Instructions.

(1)	(2)	(3)	(4)	(5)	(6)	(7)	(8)
Subgroup Number	Subgroup Size	Subgroup Defectives	p = (3)÷(2)	sqrt of (2)	C ÷ (5)	LCL = pbar-(6)	UCL = pbar+(6)
1							
2							
3							
4							
5							
6							
7							
8							
9							
10							
11							
12							
13							
14							
15							
16							
17							
18							
19							
20							
totals							

Figure 7.18 *(Continued)*

Example. Bob, our friendly neighborhood produce manager, has kept track of the number of bruised peaches in recent shipments. The following data were obtained by opening randomly selected crates from each shipment and counting the number of bruised peaches (see Figures 7.19 and 7.20). There are 250 peaches per crate. Normally Bob samples only one crate per shipment. However, when his part-time helper is available Bob samples two crates.

Shipment	Crates checked	Peaches checked	Bruised peaches
1	1	250	47
2	1	250	42
3	1	250	55
4	1	250	51
5	1	250	46
6	1	250	61
7	1	250	39
8	1	250	44
9	1	250	41
10	1	250	51
11	2	500	88
12	2	500	101
13	2	500	101
14	1	250	40
15	1	250	48
16	1	250	47
17	1	250	50
18	1	250	48
19	1	250	57
20	1	250	45
21	1	250	43
22	2	500	105
23	2	500	98
24	2	500	100
25	2	500	96

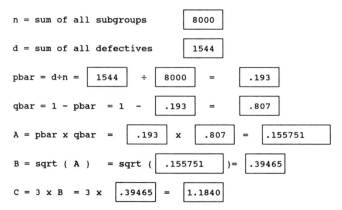

```
n = sum of all subgroups        8000

d = sum of all defectives       1544

pbar = d÷n =   1544    ÷   8000    =    .193

qbar = 1 - pbar  = 1  -    .193    =    .807

A = pbar x qbar   =    .193   x   .807   =  .155751

B = sqrt ( A )   = sqrt (  .155751  )=  .39465

C = 3 x B  = 3 x   .39465   =   1.1840
```

Figure 7.19 Example of completed *p*-chart worksheet.

(1)	(2)	(3)	(4)	(5)	(6)	(7)	(8)
Subgroup Number	Subgroup Size	Subgroup Defectives	p = (3)÷(2)	sqrt of (2)	C ÷ (5)	LCL = pbar-(6)	UCL = pbar+(6)
1	250	47	.188	15.8114	.07488	.1181	.2679
2	250	42	.168	"	"	"	"
3	250	55	.220	"	"	"	"
4	250	51	.204	"	"	"	"
5	250	46	.184	"	"	"	"
6	250	61	.244	"	"	"	"
7	250	39	.156	"	"	"	"
8	250	44	.176	"	"	"	"
9	250	41	.164	"	"	"	"
10	250	51	.204	"	"	"	"
11	500	88	.176	22.3607	.05295	.1400	.2460
12	500	101	.202	"	"	"	"
13	500	101	.202	"	"	"	"
14	250	40	.160	15.8114	.07488	.1181	.2679
15	250	48	.192	"	"	"	"
16	250	47	.188	"	"	"	"
17	250	50	.200	"	"	"	"
18	250	48	.192	"	"	"	"
19	250	57	.228	"	"	"	"
20	250	45	.180	"	"	"	"
21	250	43	.172	"	"	"	"
22	500	105	.210	22.3607	.05295	.1400	.2460
23	500	98	.196	"	"	"	"
24	500	100	.200	"	"	"	"
25	500	96	.192	"	"	"	"
totals	8000	1544					

Figure 7.19 (*Continued*)

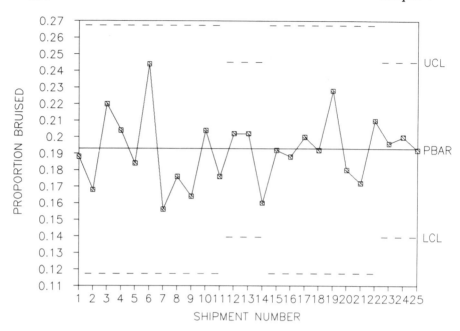

Figure 7.20   Example of completed *p* chart.

5. Pointers for real-world use. Determine if "moving control limits" are really necessary. It may be possible to use the average sample size (total number inspected divided by number of subgroups) to calculate control limits. For instance, in the example above the sample size doubled from 250 to 500 peaches but the control limits hardly changed at all. The following table illustrates the different control limits based on 250 peaches, 500 peaches, and the average sample size, which is 8000/25 = 320 peaches.

Sample size	Lower control limit	Upper control limit
250	.1181	.2679
500	.1400	.2460
320	.1268	.2592

Notice that our conclusions regarding process performance are the same when using the average sample size as when using the exact sample sizes. This is usually the case if the variation in sample size isn't too great. There are many rules of thumb, but most of them are extremely conservative. The best way to evaluate limits based on the average sample size is to check it out the way we did here.

Other SPC Techniques

A wide variety of other SPC techniques exist. A complete discussion of these techniques is beyond the scope of this book. See Pyzdek (1989) for some examples.

Process Capability Analysis

Definition. Process capability analysis is a two-stage process that involves (1) bringing a process into a state of statistical control for a reasonable period of time and (2) comparing the long-term process performance to management or engineering requirements.

Usage. Process capability analysis can be done with either attribute data or continuous data *if and only if the process is in statistical control* and has been for a reasonable period of time. Two variations on traditional process capability analysis allow use of the statistical approach with limited data: the process potential study, where only 30 consecutive units are checked (Ford Motor Company), and the mini-capability study, where only 10 consecutive units are checked (General Motors). Application of process capability methods to processes that are not in statistical control results in unreliable estimates of process capability and should never be done. Figure 7.21 illustrates these concepts.

*How to Perform Process Capability Studies, a 10-Step Plan**

1. Select a candidate for the study. This step should be institutionalized. A goal of any organization should be ongoing process improvement. However, because a company has only a limited resource base and can't solve all problems simultaneously, it must set priorities

---

PROCESS CAPABILITY

UPPER SPECIFICATION LIMIT

IN CONTROL AND CAPABLE
(VARIATION FROM COMMON CAUSES
REDUCED)

TIME

LOWER SPECIFICATION LIMIT

IN CONTROL BUT NOT CAPABLE
(VARIATION FROM COMMON CAUSES
EXCESSIVE)

SIZE

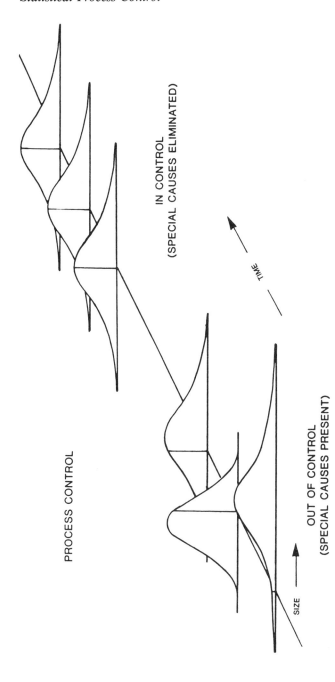

Figure 7.21   Process control concepts illustrated. (From *Continuing Process Control and Process Capability Improvement,* p. 4a. Copyright 1983. Used by permission of the publisher, Ford Motor Company, Dearborn, Michigan.)

for its efforts. The tools for this include Pareto analysis and fishbone diagrams.

2. Define the process. It is all too easy to slip into the trap of solving the wrong problem. Once the candidate area has been selected in step 1, define the scope of the study. A process is a unique combination of machines, tools, methods, and personnel engaged in production. Each element of the process should be identified at this stage. This is not a trivial exercise. The input of a large number of people may be required. There are likely to be a number of conflicting opinions about what the process actually involves. These differences must be resolved into a consensus. Finally, when the consensus has been achieved, a flowchart of the process must be prepared.

3. Procure resources. Process capability studies disrupt production and require significant expenditures of both material and human resources. Since this is a project of major importance, it should be managed as such. All of the usual project management techniques should be brought to bear. This includes planning, scheduling, and management status reporting.

4. Evaluate the measurement system. Using the SPC techniques, evaluate the measurement system's ability to do the job. Again, be prepared to spend the time necessary to get a valid means of measuring the process before going ahead.

5. Provide a control system. The purpose of the control system is twofold: (1) isolate and control as many important variables as possible, and (2) provide a mechanism for tracking variables that cannot be adequately controlled. The object of the capability analysis is to determine what the process can do if it is operated as it should be. This means that such obvious sources of potential variation as operators and vendors will be controlled while the study is conducted. In other words, a single well-trained operator will be used and the material will be from a single vendor.

There are usually some variables that are important but are not controllable. One example is the ambient environment, such as temperature, barometric pressure, or humidity. Certain process variables may degrade as part of the normal operation; for example, tools wear and chemicals are used. These variables should still be tracked using logsheets and similar tools.

6. Select an SPC method to use for the analysis. The SPC method will depend on the decisions made up to this point. If the performance

measure is an attribute, one of the attribute charts will be used. Variables charts will be used for process performance measures assessed on a continuous scale. Also considered will be the skill level of the personnel involved, need for sensitivity, and other resources required to collect, record, and analyze the data.

7. Gather and analyze the data. This entire chapter is devoted to techniques to help you do this. It is usually advisable to have at least two people go over the data analysis to catch inadvertent errors in transcribing data or performing the analysis.

8. Track down and remove special causes of variation. A special cause of variation may be obvious, or it may take months of painstaking investigation to find it. The effect of the special cause may be good or bad. Removing a special cause that has a bad effect usually involves eliminating the cause itself. For example, if poorly trained operators are causing variability, the special cause is the training system (not the operator) and it is eliminated by developing an improved training system or a process that requires less training. However, the removal of a beneficial special cause may actually involve incorporating the special cause into the normal operating procedure. For example, if it is discovered that materials with a particular chemistry produce a better product, the special cause is the newly discovered material and it can be made a common cause simply by changing the specification to assure that the new chemistry is always used.

9. Estimate process capability. One point cannot be overemphasized: the process capability cannot be estimated until a state of statistical control has been achieved. After this stage has been reached, the methods described in the next section may be used. After the numerical estimate of process capability has been arrived at, it must be compared to management's goals for the process, or it can be used as an input into economic models. Deming (1982) describes a simple economical model that can be used to determine if the output from a process should be sorted 100% or shipped as is.

10. Establish a quality improvement and control plan for the process. Once a stable process state has been attained steps should be taken to maintain it and, hopefully, improve on it. Statistical process control is one means of doing this. Others include employee involvement programs, quality circles, and quality of work life programs. Far more important than the particular approach taken is a company environment that makes continuous improvement a normal part of the daily routine.

Statistical Analysis of Process Capability Data

Control Chart Method: Attributes Data

1. Collect samples from 25 or more subgroups of consecutively produced units. Use the guidelines presented in the appropriate control chart section to determine the subgroup size. Follow the guidelines presented in the section on process capability studies for setting up and conducting the study.

2. Plot the results on the appropriate control chart (e.g., *p* chart). If all groups are in statistical control, go to the next step. Otherwise identify the special cause of variation, take action to eliminate it, and repeat step 1.

3. Using the control limits from the previous step (called operation control limits), put the control chart to use for a period of time. Once you are satisfied that sufficient time has passed for most special causes to have been identified and eliminated, as verified by the control charts, go to the next step.

4. The process capability is estimated as the control chart *centerline*. The centerline on attribute charts is the long-term expected quality level of the process (e.g., the average proportion defective). This is the level created by the common causes of process variation.

5. If the process capability doesn't meet management requirements, take immediate action to modify the process for the better. Whether it meets requirements or not, always be on the lookout for possible process improvements. The control charts will provide verification of improvement.

Control Chart Method: Variables Data

1. Collect samples from 25 or more subgroups of consecutively produced units, following the 10-step plan described above.

2. Plot the results on the appropriate control chart (e.g., *X*-bar and *R* chart). If all groups are in statistical control, go to the next step. Otherwise identify the special cause of variation, take action to eliminate it, and repeat step 1.

3. Using the control limits from the previous step (called operation control limits), put the control chart to use for a period of time. Once you are satisfied that sufficient time has passed for most special causes to have been identified and eliminated, as verified by the control charts, go to the next step.

4. The process capability is estimated from the process average and standard deviation. The standard deviation is computed based on the average range, in the case of individual charts the average difference is used. This is the level of variability created by the common causes of process. The formula is

$$\hat{\sigma} = \frac{\bar{R}}{d_2} \qquad \text{[Equation (20) repeated]}$$

5. The value d_2 was discussed earlier. It is a constant that is obtained from Table 6 in the Appendix.

6. Only now can the process be compared to engineering requirements. One way of doing this is by calculating "capability indexes." Following are several popular capability indexes and their interpretation.

Process Capability Indexes

$$C_p = \frac{\text{engineering tolerance}}{6\hat{\sigma}} \tag{29}$$

$$C_R = 100 \times \frac{6\hat{\sigma}}{\text{engineering tolerance}} \tag{30}$$

$$Z_U = \frac{\text{upper specification} - \text{process average}}{\hat{\sigma}} \tag{31}$$

$$Z_L = \frac{\text{process average} - \text{lower specification}}{\hat{\sigma}} \tag{32}$$

$$Z_{\min} = \text{Minimum}\{Z_L, Z_U\} \tag{33}$$

$$C_{pk} = \frac{Z_{\min}}{3} \tag{34}$$

These capability indexes are listed below with interpretations and comments.

C_p This is the most common capability index. It simply makes a direct comprison of the process spread to the engineering requirements. Assuming the process distribution is normal and the process average is exactly centered between the engineering requirements, a C_p index of 1 would give a "capable process." However, to allow a bit of room for process drift, the generally accepted minimum value for C_p is 1.33. This situation is shown in Figure 7.22. In general, the larger C_p is, the better. The C_p index has two major shortcomings: it can't be used unless there are both upper and lower specifications, and it does not account for process centering. If the process average is not exactly centered relative to the engineering requirements, the C_p index will give misleading results.

C_R The C_R index is exactly equivalent to the C_p index. It simply makes a direct comparison of the process to the engineering requirements. Assuming the process distribution is normal and the process average is exactly centered between the engineering requirements, a C_R index of 100 would give a "capable process." However, to allow a bit of room for process drift, the generally accepted maximum value for C_R is 75. In general, the smaller C_R is, the better. The C_R index suffers from the same shortcomings as the C_p index.

Z_U The Z_U index measures the process location (central tendency) relative to its standard deviation and the upper requirement. If the distribution is normal, the value of Z_U can be used to determine the percentage above the upper requirement by using

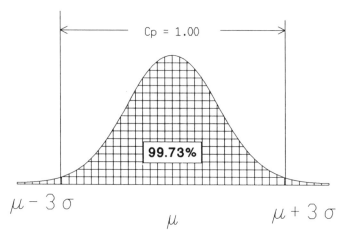

Figure 7.22 Illustration of a capable process.

Table 1 in Appendix. The method is the same as described in the previous chapter using the Z transformation; simply use Z_U instead of Z. In general, the bigger Z_U is, the better. A value of at least $+3$ is required to assure that 0.1% or less defective will be produced. A value of $+4$ is generally desired to allow some room for process drift.

Z_L The Z_L index measures the process location relative to its standard deviation and the lower requirement. If the distribution is normal, the value of Z_L can be used to determine the percentage above the upper requirement by using Table 1 in the Appendix. The method is the same as described in the previous chapter using the Z transformation, except that you use $-Z_L$ instead of Z. In general the bigger Z_L is, the better. A value of at least $+3$ is required to assure that 0.1% or less defective will be produced. A value of $+4$ is generally desired to allow some room for process drift.

Z_{min} The value of Z_{min} is simply the smaller of the Z_L and Z_U values. It can be considered an intermediate result to be used in computing C_{pk}.

C_{pk} The value of C_{pk} is simply Z_{min} divided by 3. Since the smallest value represents the nearest specification, the value of C_{pk} tells you if the process is truly capable of meeting requirements. A C_{pk} of at least $+1$ is required, and $+1.33$ is preferred. Note that C_{pk} is closely related to C_p; the difference between C_{pk} and C_p represents the potential gain to be had from centering the process.

Example of Capability Analysis Using Normally Distributed Variables Data. Assume we have conducted a capability analysis using X-bar and R charts with subgroups of 5. Also assume that we found the process to be in statistical control with a grand average of 0.9983 and an average range of 0.02205. From the table of d_2 values (Appendix Table 6) we find d_2 is 2.326 for subgroups of 5. Thus, using equation (20).

$$\hat{\sigma} = \frac{0.02205}{2.326} = 0.00948$$

Before we can analyze process capability, we must know what the requirements are. For this process the requirements are a lower specification of 0.980 and an upper specification of 1.020 (1.000 plus or minus 0.020). With this information, plus the knowledge that the process performance has been in statistical control, we can compute the capability indexes for this process.

$$C_p = \frac{1.020 - 0.9800}{6 \times 0.00948} = \frac{0.04}{0.05688} = 0.703$$

$$C_R = 100 \times \frac{0.05688}{0.04} = 142.20\%$$

$$Z_U = \frac{1.020 - 0.99832}{0.00948} = \frac{0.02168}{0.00948} = +2.3$$

$$Z_L = \frac{0.99832 - 0.980}{0.00948} = \frac{0.01832}{0.00948} = +1.9$$

$$Z_{min} = \text{Minimum}\{1.9, 2.3\} = 1.9$$

$$C_{pk} = \frac{1.9}{3} = 0.63$$

OK, so what does all of this mean? The answer is found by interpreting the capability indexes. Let's look at them one at a time.

Index	Value	Interpretation
C_p	0.703	Since the minimum acceptable value for this index is 1, the result 0.703 indicates that this process cannot meet the requirements. Furthermore, since the C_p index doesn't consider process centering, we know that the process can't be made acceptable by merely adjusting the process closer to the center of the requirements. Thus we can expect the Z_L, Z_U, and Z_{min} values to be unacceptable too.
C_R	142.2%	This value always gives the same conclusions as the C_p index. The number itself means that the "natural tolerance" of the process uses 142.2% of the engineering requirement.
Z_U	+2.3	We desire a Z_U of at least +3, so this value is unacceptable, as we predicted from the C_p index. We can use Z_U to estimate the percentage of production that will exceed the upper specification. Referring to the interpretation of Z_U above, we find that approximately 1.1% will be oversized.

Index	Value	Interpretation
Z_L	+1.9	We desire a Z_L of at least +3, so this value is unacceptable, as we predicted from the C_p index. We can use Z_L to estimate the percentage of production that will be below the lower specification. Referring to the interpretation above, we find that approximately 2.9% will be undersized. Adding this to the 1.1% of oversized, we estimate a total reject rate of 4.0%. By subtracting this term from 100% we get the projected yield of 96.0%.
Z_{min}	+1.9	Obviously, this value will not be acceptable.
C_{pk}	0.63	The value of C_{pk} is only slightly smaller than that of C_p. This indicates that we will not gain much by centering the process. The actual amount we would gain can be calculated by "assuming" the process is exactly centered at 1.000 and recalculating Z_{min}; if we do this we find the predicted total reject rate drops from 4.0 to 3.6%.

SUMMARY

This chapter introduced the concept of statistical control. Statistical process control was defined. Variation as a fact of life was discussed, along with ways of characterizing variation in a process. The central limit theorem was presented (without proof) and its importance in SPC was explained. The differences between prevention-based control systems and systems based on detection were presented. The SPC techniques of Pareto analysis, cause-and-effect diagrams, histograms, X-bar and range charts, p charts, and process capability analysis were introduced along with worked-out examples.

This chapter did not cover the relationships among the various SPC techniques, and the application of the SPC techniques to troubleshooting was not discussed (e.g., the patterns that appear on the histograms and control charts can often be related to causes listed on cause-and-effect diagrams). Omitted were various additional control charts for both attributes and variables data. Also omitted were a number of alternatives to traditional control charting for process control, such as analysis of variance (ANOVA) and chi-square tests. Finally, process capability analysis of nonnormally distributed data was not discussed.

8

New Quality Technologies

Quality is becoming increasingly important as businesses grapple for a competitive advantage. A host of new methods and techniques have been developed in response to the need for expanding levels of quality. Some of these new technologies are not really new; for example, Taguchi methods have existed for over four decades. However, the level of interest is certainly without precedent.

This chapter provides a brief introduction to several popular quality technologies. The brevity is necessary because of the vast amount of material being covered. Topics include statistical design of experiments, Taguchi methods, and quality function deployment (QFD).

STATISTICAL DESIGN OF EXPERIMENTS

Designed experiments play an important role in quality improvement. This chapter will introduce the basic concepts involved and it will contrast the statistically designed experiment with the "one variable at a time" approach that has been used traditionally. Readers interested in this subject are referred to Taguchi (1987) and Montgomery (1984).

TERMINOLOGY

Designed experiment. An experiment where one or more variables (called independent variables) that are believed to have an effect on the experimental outcome are identified and manipulated according to a predetermined plan. Data collected from a designed experiment can be analyzed statistically to determine the effect of an independent variable or a combination of independent variables. An experimental plan must also include provisions for dealing with extraneous variables, that is, variables not explicitly identified as independent variables.

Response variable. The variable being investigated, also called the dependent variable.

Primary variables. The controllable variables believed most likely to have an effect. These may be quantitative, such as temperature, pressure, or speed, or qualitative, such as vendor, production method, or operator.

Background variables. Variables, identified by the designers of the experiment, which may have an effect but either cannot or should not be deliberately manipulated or held constant. The effect of background variables can contaminate primary-variable effects unless they are properly handled. The most common method of handling background variables is blocking (described later in this chapter).

"Common causes" or experimental error. In any given experimental situation, a great many variables may be potential sources of variation—so many, in fact, that no experiment could be designed that deals explicitly with every possible source of variation. The variables that are not dealt with explicitly are analogous to the common causes of variation described in Chapter 7. They represent the "noise level" of the process, and their effects are kept from contaminating the primary-variable effects by randomization. Randomization is a procedure that assigns test units to test conditions in such a way that any given unit has the same probability as any other unit of being processed under a given set of test conditions.

Fixed-effects model. An experimental model in which all factor levels of interest are studied. For example, if there are three different materials, all three are included in the experiment.

Random-effects model. Experimental model in which the levels of factors evaluated by the experiment represent a sample of all possible levels—for example, if we have three different materials but only use two materials in the experiment.

Mixed model. Experimental model with both fixed and random factors.

Completely randomized design. Experimental plan where the order in which the experiment is performed is completely random. For example:

Level	Test sequence number
A	7,1,5
B	2,3,6
C	8,4

Randomized block design. Experimental design in which the experimental observations are divided into "blocks" according to some criterion. The blocks are filled sequentially, but the order within each block is filled randomly. For example, assume we are conducting a painting test with different materials, material A and material B. We have four test pieces of each material. Ideally, we would like to clean all of the peices at the same time to assure that the cleaning process doesn't have an effect on our results. But what if our test requires that we use a cleaning tank that cleans two test pieces at a time? The tank load then becomes a "blocking factor." We will have four blocks, which might look like this:

Material	Tank load	Test piece number
A	1	7
B		1
B	2	5
A		2
B	3	3
A		6
B	4	4
A		8

Since each material appears exactly once per cleaning tank load, the material totals or averages can be compared directly. The reader should be aware that statistical designs exist to handle more complicated "unbalanced designs" (for details see Montgomery, 1984).

Replication. Collection of more than one observation for the same set of experimental conditions.

Interaction. Condition in which the effect of one factor depends on the level of another factor. Interaction is illustrated in Figure 8.1. The chart on the left illustrates a situation where the effect of varying factor A depends on the level of factor B. If B is at its low level, then increasing factor A from its low level to its high level results in increasing the response variable. However, increasing factor A when factor B is at its high level has the opposite effect. The chart on the right illustrates no interaction, in that varying factor A has the same effect regardless of the level of factor B.

THE "TRADITIONAL APPROACH" VERSUS STATISTICALLY DESIGNED EXPERIMENTS

The traditional approach is to hold all variables except one constant. The statistically designed experiment usually involves varying two or

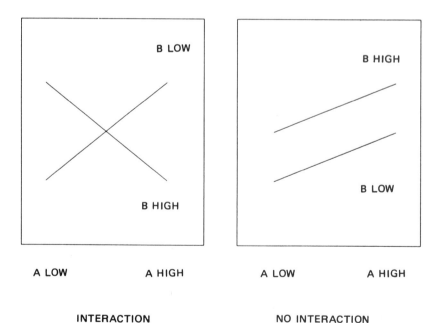

Figure 8.1 Illustration of interaction.

more variables simultaneously. The advantages of the statistical approach are twofold:

1. Interactions, such as those illustrated above, can be detected and measured. Failure to detect interactions is a major flaw in the "one variable at a time" approach.
2. Each value does the work of several values. A properly designed experiment allows you to use the same observation to estimate several different effects. This translates directly to cost savings when using the statistical approach.

TAGUCHI METHODS

This section presents some of the special concepts introduced by Dr. Genichi Taguchi of Japan. The reader is referred to Taguchi (1987) for a complete discussion of his approach to quality engineering. Many of Taguchi's ideas are useful in that they present an alternative way of looking at quality in general; however, they are controversial.

Quality is defined as the loss imparted to the society from the time a product is shipped (Taguchi, 1986). Taguchi divides quality control efforts into two categories: on-line quality control and off-line quality control.

On-line quality control involves diagnosis and adjustment of the process, forecasting and correction of problems, inspection and disposition of product, and follow-up on defectives shipped to the customer.

Off-line quality control methods are quality and cost control activities conducted at the product and process design stages in the product development cycle. Kackar (1985) lists three major aspects of off-line quality control:

System design: the process of applying scientific and engineering knowledge to produce a basic functional prototype design. The prototype model defines the initial settings of product or process design characteristics.
Parameter design: an investigation conducted to identify settings that minimize (or at least reduce) the performance variation. A product or a process can perform its intended function at many settings of its design characteristics. However, variation in the performance characteristics may change with different settings. This variation increases both product manufacturing

and lifetime costs. The term parameter design comes from an engineering tradition of referring to product characteristics as product parameters. An exercise to identify optimal parameter settings is therefore called parameter design.

Tolerance design: a method for determining tolerances that minimize the sum of product manufacturing and lifetime costs. The final step in specifying product and process designs is to determine tolerances around the nominal settings identified by parameter design. It is still a common practice in industry to assign tolerances by convention rather than scientifically. Tolerances that are too narrow increase manufacturing costs, and tolerances that are too wide increase performance variation and the lifetime cost of the product.

Measurement

Figure 8.2 Quadratic loss function.

Expected loss refers to the monetary losses an arbitrary user of the product is likely to suffer at an arbitrary time during the product's life span because of performance variation. Taguchi advocates modeling the loss function so that the issue of parameter design can be made more concrete. The most often used model of loss is the quadratic loss function illustrated in Figure 8.2.

The traditional American model of loss is illustrated in Figure 8.3. Note the contrast between the quadratic loss function and the conceptual loss function traditional in America.

The interpretation of Figure 8.3 is that there is no loss as long as product meet requirements. A part that is just barely inside specifications is as good as one that is exactly nominal. A part that is beyond the specification limit incurs a loss. Parts that deviate a great deal from specifications incur the same loss as those just barely outside (usually scrap or rework).

Noise is the term used to describe all the variables, except design parameters, that cause performance variation during a product's life

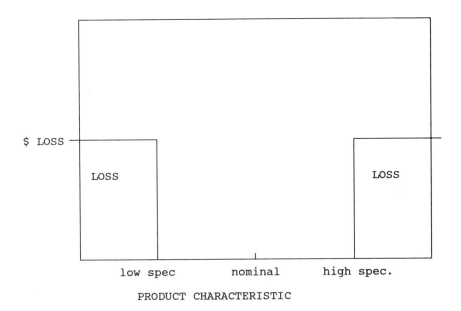

Figure 8.3 Traditional American view of loss (goalpost model).

span and across different units of the product. Sources of noise are classified as either external sources or internal sources.

> *External sources of noise* are variables external to a product that affect the product's performance.
>
> *Internal sources of noise* are the deviations of the actual characteristics of a manufactured product from the corresponding nominal settings.
>
> *Performance statistics* estimate the effect of noise factors on the performance characteristics. Performance statistics are chosen so that maximizing the performance measure will minimize expected loss. Many performance statistics used by Taguchi use "signal to noise ratios" which account jointly for the levels of the parameters and the variation of the parameters.

Summary of the Taguchi Method

The Taguchi method for identifying settings of design parameters that maximize a performance statistic is summarized by Kackar (1985) as follows:

> Identify initial and competing settings of the design parameters, and identify important noise factors and their ranges.
>
> Construct the design and noise matrices, and plan the parameter design experiment.
>
> Conduct the parameter design experiment and evaluate the performance statistic for each test run of the design matrix.
>
> Use the values of the performance statistic to predict new settings of the design parameters.
>
> Confirm that the new settings indeed improve the performance statistic.

What Is a Taguchi Experiment?

There is a great deal of confusion over what is a Taguchi experiment and what is not. A paper by Yuin Wu of the American Supplier Institute describes Taguchi experiments as follows:

1. Best selection of quality characteristics (go/no-go is not good)
2. Maximum possible number of control factors
3. Comparison of existing conditions with predicted optimum conditions

4. Use of signal-to-noise ratios
5. Use of loss function
6. Minimum interactions among control factors
7. Control factor and noise factor separated
8. Use of orthogonal arrays

QFD

QFD is a system for design of a product or service based on customer demands, a system that moves methodically from customer requirements to requirements for the products or services. QFD is also an approach that involves the entire company in the design and control activity. Finally, QFD provides documentation for the decision-making process. The QFD approach involves four distinct phases (King, 1987):

Organization phase. Management selects the product or service to to be improved, the appropriate interdepartmental team, and defines the focus of the QFD study.

Descriptive phase. The team defines the product or service from several different directions such as customer demands, functions, parts, reliability, cost, and so on.

Breakthrough phase. The team selects areas for improvement and finds ways to make them better through new technology, new concepts, better reliability, cost reduction, etc., and monitors the bottleneck process.

Implementation phase. The team defines the new product and how it will be manufactured.

QFD is implemented through the development of a series of matrices. In its simplest form QFD involves a matrix that presents customer requirements as rows and product or service requirements as columns. The cells, where the rows and columns intersect, show the correlation between the individual customer requirement and the product or service requirement. This is called the "A-1 matrix" or the "requirement matrix." Figure 8.4 shows the layout of a requirements matrix.

QFD proceeds by developing a series of four related matrices (King, 1987): product planning matrix, part deployment matrix, process planning matrix, and production planning matrix. Each matrix is

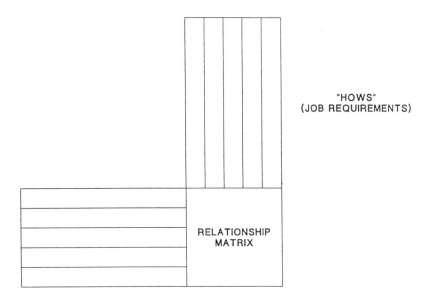

"HOWS"
(JOB REQUIREMENTS)

RELATIONSHIP
MATRIX

"WHATS"
(CUSTOMER REQUIREMENTS)

Figure 8.4   Requirements matrix layout.

related to the previous matrix as shown in Figure 8.5.

The requirements matrix matrix develops what is called the "house of quality." Figure 8.6 illustrates one approach (Fortuna, 1988).

The house of quality relates, in a simple graph format, customer requirements, product characteristics, and competitive analysis. It is crucial that this matrix be developed carefully and completely since it forms the basis of the entire QFD process. By following the QFD approach, the customer's demands are "deployed" to the final process and product requirements.

This implementation of QFD is a hybrid of work done by two Japanese pioneers in QFD, Macabe and Fukahara. An alternative approach to QFD, based on the work of Yoji Akao, is described by King (1987).

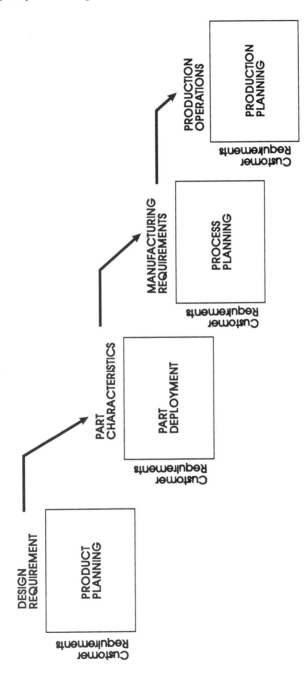

Figure 8.5 QFD methodology, Macabe approach (King, 1987).

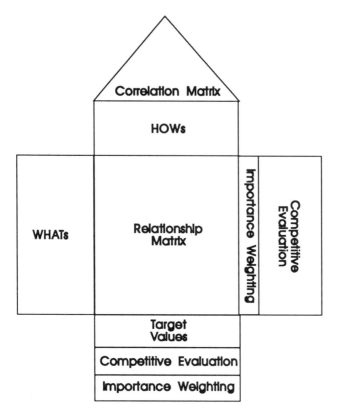

Figure 8.6   The house of quality. (From Fortuna, 1988, pp. 23–28).

THE SEVEN-M-TOOLS

Since Dr. Shewhart launched modern quality control technology in 1931 with his book *Economic Control of Quality of Manufactured Product,* the literature of quality control technology has been changing constantly. In recent years, the pace of change of quality technology has been accelerating. The seven-M-tools are an example of this change. This section will provide definitions of the seven-M-tools. The reader is referred to Mizuno (1988) for additional information on these techniques.

1. Affinity diagram. A means of organizing ideas into significant categories. These ideas are often generated in brainstorming sessions. The technique works as follows:

1. Write the ideas on small pieces of paper (e.g., Post-its™).
2. The team works *in silence* to arrange the ideas into separate categories. To put an idea into a category a person simply picks up the Post-it™ and moves it.

By working in silence the right side of the brain is involved. Since the right side of the brain excels at pattern recognition, this approach is thought to have merit. The final groupings are then reviewed and discussed by the team. Usually, the grouping of ideas helps develop a coherent plan.

Affinity diagrams are useful for analysis of quality problems, defect data, customer complaints, survey results, and so on. These diagrams can be used in conjunction with other techniques such as cause-and-effect diagrams (see Chapter 7) or interrelationship digraphs. Figure 8.7 presents a sample affinity diagram.

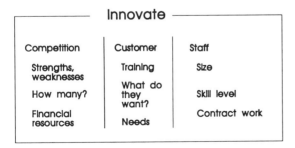

Figure 8.7   Affinity diagram.

2. Interrelationship digraph. Like affinity diagrams, interrelationship digraphs are designed as a means of organizing disparate ideas, usually (but not always) generated in brainstorming sessions. However, whereas affinity diagrams seek to simply arrange ideas into groups, interrelationship digraphs attempt to define the ways in which ideas influence each other. It is best to use *both* affinity diagrams and interrelationship digraphs.

The interrelationship digraph begins by writing the ideas on small pieces of paper, such as Post-its™. These pieces of paper are then placed on a large sheet of paper, such as a flipchart or a piece of large blueprint paper. Arrows are drawn between related ideas. An idea that has arrows leaving but not entering is a "root idea." By evaluating the relationships between ideas you will get a better picture of the way things happen. The root ideas are often keys to improving the system. Figure 8.8 shows a basic interrelationship digraph.

3. Tree diagram. Used to break down or stratify ideas in progressively greater detail. The objective is to partition a big idea or problem into its components. By doing this you will make the idea easier to understand or the problem easier to solve. Figure 8.9 shows an example of a tree diagram.

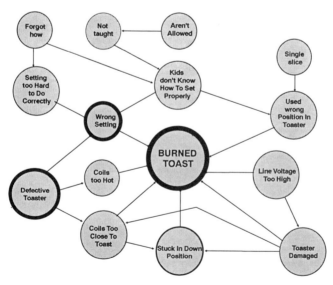

Figure 8.8 Interrelationship digraph. How does toast get burned?

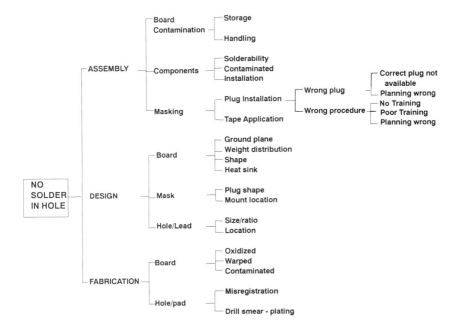

Figure 8.9   Example of a tree diagram.

4. Matrix chart. This chart is constructed to analyze the correlations between two groups of ideas. Actually, QFD is a series of matrix charts. The major advantage of constructing matrix charts is that it forces you to *systematically* analyze correlations. Matrix charts can be used in conjunction with decision trees. To do this, simply use the most detailed level of two decision trees as the contents of rows and columns of a matrix chart.

5. Matrix data analysis chart. Used to portray the correlations of multiple variables to selected important factors. The layout of the matrix data analysis chart consists of creating two lines that cross each other at right angles. Each line represents a different criterion or factor. Multiple concepts are then plotted as they correlate to the two criteria. The intersection of the two lines is zero correlation. Positive correlation with criterion #1 lies to the right of the intersection, i.e., as criterion #1 increases it has a positive impact on the concept. Negative correlation with criterion #1 lies to the left of the intersection, i.e., as criterion #1 increases it has a negative impact on the concept. Positive correlation with criterion #2 lies above the intersection, negative correlation with criterion #2 lies below the intersection.

The matrix data analysis chart may have its roots in factor analysis, a mathematical method of analyzing the interrelationships of multiple variables. As with factor analysis, matrix data analysis attempts to reduce the problems to a relatively small number of important criteria. The idea is that it is easier to focus on the smaller number of criteria than the larger number of concepts. Figure 8.10 presents a matrix data analysis chart.

6. *Process decision program chart (PDPC).* A technique designed to help prepare contingency plans. It is modeled after reliability engineering methods such as failure mode, effects, and criticality analysis (FMECA), and fault tree analysis. The emphasis of PDPC is the impact of "failures" (problems) on project schedules. Also, PDPC describes specific actions taken to mitigate the impact of the problems. To en-

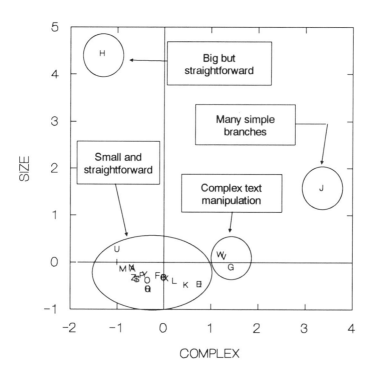

Figure 8.10 Matrix data analysis chart.

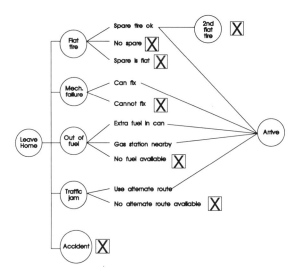

Figure 8.11 Process decision program chart.

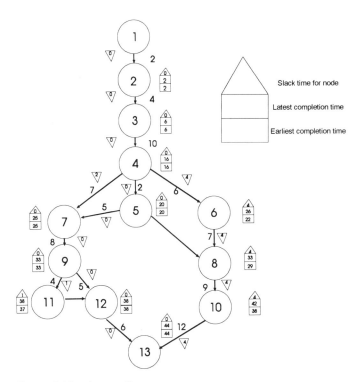

Figure 8.12 Arrow diagram.

hance classical PDPC, assign subjective probabilities to the problems and use these to help assign priorities. Figure 8.11 shows a process decision program chart.

7. *Arrow diagram.* Like many other "new" tools, the arrow diagram has its roots in well-established methods used in another discipline, in this case, operations research. The arrow diagram is directly analagous to the critical path method (CPM) and the program evaluation and review technique (PERT). These two project management tools have been used for many years to determine which activities must be performed, when they must be performed, and in what order. Unlike CPM and PERT, which require advanced education in project management or systems engineering, arrow diagrams are greatly simplified so that they can be used with a minimum of training. Figure 8.12 shows an arrow diagram.

SUMMARY

The advantages of statistical design over the classical one-at-a-time approach were presented. The basic concepts and definitions of experimental design were introduced. The fundamental concepts of Taguchi's method of quality control were discussed. A brief description of QFD was presented. Finally, thumbnail descriptions of the Seven-M-Tools were provided.

The chapter did not cover the mathematics of experimental design or Taguchi methods. No formulas were provided, nor was any detailed framework provided for conducting a designed experiment or analyzing the data from one.

9

The Modern Approach to Quality Management

INTRODUCTION

The last chapter presented some of the emerging new quality technologies. However, there is a great deal more than technology involved in reaching world-class quality levels. By now it is no secret that the Japanese are major-league competitors in the international marketplace. Equally well known is the fact that this position came about partly as a result of the superior quality of Japanese goods. Since this superiority encompasses a broad range of products, from microelectronic devices to heavy industrial machinery, it seems likely that Japanese manufacturers as a group are doing something noteworthy. This chapter explores the way in which the Japanese approach quality.

Perhaps a starting point is to examine history. It is important to remember that the Japanese were engaged in manufacturing in a significant way well before World War II, as evidenced by their arsenal of weapons. Their fleet of aircraft carriers and airplanes attest to that. However, the reality in postwar Japan was that their quality levels were too poor to give them much access to international markets. The American occupation forces understood the significance of export markets to Japan, which is only a small island with 110 million people.

Clearly, quality improvement had to be one of Japan's highest priorities.

Actually, the Japanese had begun their education somewhat before the American occupation. During World War II the Japanese were plagued by quality problems. Part of their wartime quality improvement involved translating works on statistical quality control methods into Japanese. However, the real push for implementing modern quality control methods came with a series of seminars conducted in 1950 by Dr. W. Edwards Deming, an American statistician and quality expert. Perhaps more important than the technical tools, Deming's philosophy of management was imparted to Japanese top management. The wholehearted endorsement of Deming's philosophy by Japanese top management was probably the single most important factor in Japan's turnaround in quality. With top management as champions, the adoption of Deming's philosophy was assured. The Japanese have honored Deming by naming their most prestigious industrial award in his honor: the Deming award.

Deming's philosophy has been embodied in his 14 Points for Management, which are listed in condensed form in Table 9.1. These 14 points are elaborated on in a number of different books; see Deming (1982), Scherkenbach (1986), and Gitlow and Gitlow (1987).

Other eminent American quality experts also made important contributions to Japanese quality improvement efforts. Our overall approach to quality management, including the use of statistical methods, was taught to eager Japanese audiences throughout the 1950s and 1960s. In addition to SPC, the Japanese learned of and adopted total quality control and zero defects programs. In return, we learned a great deal about quality improvement from the Japanese. Table 9.2 summarizes the exchange of information between American and Japan.

As Table 9.2 shows, the information exchange has been a two-way street. Many authors and consultants seem to believe that all useful QC methods are the property of American genius, expropriated unfairly by the Japanese. This is simply not the case. The free exchange of ideas and information has been a tradition in both American and Japan.

DIFFERENCES

As a quality professional, I have been interested in the Japanese for quite some time. The quality literature first began to mention Japan in the 1960s. By the mid-1970s Japan's high levels of quality began to draw

Table 9.1 Deming's 14 Points for Management

1. Innovate and allocate resources to fulfill the long-range needs of the company and customer rather than short-term profitability.
2. Discard the old philosophy of accepting defective products.
3. Eliminate dependence on mass inspection for quality control; instead, depend on process control through statistical methods.
4. Reduce the number of multiple source suppliers. Price has no meaning without an integral consideration for quality. Encourage suppliers to use statistical process control.
5. Use statistical techniques to identify the two sources of waste—system (85%) and local (15%) faults; strive to constantly reduce this waste.
6. Institute more thorough, better job-related training.
7. Provide supervision with knowledge of statistical methods; encourage use of these methods to identify which defects should be investigated for solution.
8. Reduce fear throughout the organization by encouraging open, two-way, nonpunitive communication.
9. Help reduce waste by encouraging design, research, and sales people to learn more about the problems of production.
10. Eliminate the use of goals and slogans to encourage productivity, unless training and management support is also provided.
11. Closely examine the impact of work standards. Do they consider quality or help anyone to do a better job? They often act as an impediment to quality.
12. Institute rudimentary statistical training on a broad scale.
13. Institute a vigorous program for retraining people in new skills, to keep up with changes in materials, methods, product designs, and machinery.
14. Make maximum use of statistical knowledge and talent in your company.

Table 9.2 American and Japanese Contributions to Quality Control

American contributions	Japanese contributions
Statistical QC methods (pre World War II)	New techniques: quality circles, fishbone diagrams, Taguchi methods
Deming's philosophy (1950)	
Juran's approach to quality management (1954)	Long-term orientation to management
Feigenbaum's total quality control approach (1961)	Externally focused management (customers, competitors, suppliers, government, society)
	Companywide involvement in QC

attention from outside the quality profession. When Japanese quality finally surpassed that of competing American goods, everyone took notice. Japanese superiority in quality and productivity became the subject of many studies. It was shown over and over again that Japanese goods were superior to their American counterparts. The list includes automobiles, videotapes, televisions, air conditioners, and a huge variety of other products.

A procession of American managers and consultants went to Japan to plumb the depths of this mystery. The result of this firsthand observation was the realization that there was nothing to see. The Japanese appeared to be using the same techniques as their American counterparts. Obviously, however, something is different. One Japanese quality expert, Dr. Kaoru Ishikawa, offered some of his observations on how the Japanese approach to quality differs from that of Western nations (Ishikawa, 1985):

1. Companywide quality control; participation by all members of the organization in quality control
2. Education and training in quality control
3. Quality control circle activities
4. Quality control audits (Deming prize audits and presidential audit)
5. Utilization of statistical methods
6. Nationwide quality control promotion activities

There are also differences in how American managers and Japannese managers define the essence of quality control. The American view is that quality has been achieved if the customer receives product that conforms to engineering requirements (i.e., zero defects). In fact, many Americans believe that an acceptable quality level (AQL) might still allow defective product to be delivered to the customer as a long-term practice. The Japanese view is that firms have a responsibility to society to reduce, and eventually eliminate, all negative impact on society caused by the firm's products and services. Quality, in the Japanese view, is defined by the customer, not by the producer.

MEASURING QUALITY

Efforts to measure quality are vital to the quality control effort. Much of this book has dealt with the subject of quality measurement. Since the importance of quality measurement is recognized by both American and Japanese firms, an examination of the way in which we measure quality might provide further clues to the differences in quality levels. Ishikawa contrasts the ways in which Japanese and Western firms express quality. He provides seven "pointers in expressing quality" (Ishikawa, 1985, pp. 49–55):

1. Determine the assurance unit (what we are to measure).
2. Determine the measuring method.
3. Determine the relative importance of quality characteristics.
4. Arrive at a consensus on defects and flaws.
5. Expose latent defects.
6. Observe quality statistically.
7. "Quality of design" and "quality of conformance."

From my experience, I believe that American manufacturers believe that they actually do all of these things. However, they are wrong. While all manufacturers may have done one or more of these things at one time or another, the application is so sporadic and haphazard that it can't be said that the *approach* was applied. The difference is crucial.

Another fundamental difference is the very way we look at quality. In America the way we think of quality is negative. Quality is measured as the absence of undesirable attributes (e.g., a unit "has quality" if it is nondefective). Ishikawa calls this *backward-looking quality*. The customer, in contrast, judges quality as the presence of desirable attributes. This is called *forward-looking quality*. The very use of the term *quality control* implies a backward-looking quality philosophy.

Even when we measure quality as the absence of undesirable attributes, we differ from the Japanese. Rework and repair are often so common that they are explicitly incorporated into manufacturing processes. On one occasion I heard a company vice-president accuse a quality improvement team of wasting their time trying to improve the quality of soldering on a circuit board, because "solder problems are never the cause of shipping problems." The team leader correctly pointed out that 40% of the cost of the circuit board was from solder touch-up and inspection after touch-up, a cost included as a normal part of the operation.

QUALITY CIRCLES

Quality circles are a Japanese contribution to quality technology. A quality circle is a small group that meets periodically to perform quality control activities. Circle membership is voluntary, and most circles are composed of people from the same work area. Circles are usually permanent and their activities are part of the broader companywide quality control effort.

Circle members are trained in the use of quality control techniques and supported by management. The idea is that all workers have an important contribution to make to the quality effort and the circle is one mechanism for allowing them the opportunity to make their contribution.

American business has adopted the quality circle approach on a massive scale, with mostly poor results. My experience has been that the circles are usually begun with great enthusiasm, produce a few significant results, and then die out. While there are many reasons for this, the most common ones are:

1. The quality circle in an American firm is isolated, not part of a companywide quality control effort. As a result, circles are usually unable to deal successfully with problems involving other areas of the company.
2. Key management personnel move about too frequently and circles are not provided with consistent leadership and management support.
3. Employees transfer in and out of circle work areas too frequently. Without stability in the membership, circles never develop into effective groups; building effective teams takes time.

VENDOR RELATIONS

The Japanese view vendors as an extension of their company. And well they should, since, on average, Japanese firms purchase 70% of the cost of their products from outside sources. Vendors are viewed as vital elements in the quality control program and are treated as such. Visits and other communications are frequent and routine. Japanese firms develop very close working relationships with a small number of suppliers. The smaller number means that each supplier can be given larger orders, which helps win their loyalty.

American companies also depend a great deal on their vendors, averaging 50% of their costs for purchased goods. However, we treat our vendors much differently than the Japanese. Multiple sources of supply are common; some government contracts even require them. Sources are pitted against each other and threatened with the loss of business. Communication is often one-sided, with the customer telling the vendor what the requirements are and informing the supplier when requirements are not satisfied. American suppliers are seldom brought in during the design phase, and their input at all stages is limited or nonexistent.

EMPLOYEE RELATIONS

I once attended an international quality conference and listened to a number of Japanese lectures. The speakers repeatedly spoke of "loving the workers" and of the workers "loving the company." In Japanese businesses the fundamental assumptions regarding employees are different than those in American firms. Many, if not most, Japanese place their loyalty to their employers above their families.

Most American businesses have paid lip service to employee involvement. Some have even invested a great deal of money in this area. The results have been mixed but for the most part unimpressive. Our history has created an atmosphere of mistrust and hostility between management and employees that is difficult to break down. Management and workers have been isolated from one another both culturally and physically. Many companies actually promote this with such things as separate office buildings, cafeterias, rest rooms, and parking lots. The solution will require more than slogans; management must demonstrate their sincerity in meaningful ways.

QUALITY AND MARKETING

The Japanese quality control effort includes aggressive pursuit of customer feedback. The marketing function plays an important role in this. This effort, combined with the forward-looking quality philosophy, actually drives the quality control effort. Customer feedback is used to assess the effectiveness of current quality control efforts and to provide the basis for improved process and product designs.

American quality control efforts make little use of marketing information, except perhaps to track warranty costs. Customer feedback is usually limited to complaints, and the complaints seldom lead to long-term corrective action. Redesign of processes or products based on customer feedback is rare.

JUST IN TIME

Many Americans have heard of just-in-time production, but few realize that it is intimately related to quality control. Just in time, or JIT, is a production goal. The objective is to operate without safety stock. The basic elements of JIT include cutting lot sizes, cutting setup times, and cutting purchase order costs. As used in Japan, JIT is part of a broader companywide effort that includes total quality control, employee involvement, and vendor involvement. Safety stock is viewed as a "narcotic drug" that hides symptoms that something is wrong and requires management attention.

American companies have been quick to adopt JIT. As often as not, it has been a disaster. The reason for this is that American companies fail to realize that JIT must be part of a companywide effort that includes total quality control, employee involvement, and vendor involvement. As stated above, safety stock is a narcotic drug that hides pain symptoms. We must complete the surgery before we remove the anesthesia!

PERFECTION VERSUS AQL

American quality control has long used a term called the acceptable quality level, or AQL. The term is defined as the maximum percentage defective that is acceptable as a long-term process average. This has become more than a mere term for many American companies: it has become an operating philosophy. "It meets the AQL" has become the battle cry for an entire generation of manufacturing managers fighting to maintain the status quo.

The Japanese philosophy is the antithesis of the AQL. It is, in a word, perfection. The concept of perfection goes beyond zero defects. It seeks as an ultimate goal the elimination of variation in quality. Perfection is, of course, unattainable. However, its value as a guiding philosophy is considerable, and its advantages over the AQL philosophy are obvious.

AMERICAN INNOVATORS

Probably the most successful American innovators to date have been the Japanese themselves! While some in this country continue to argue that the Japanese success is based on cultural differences, the fact is that Japanese management has already successfully transformed several companies in the United States. For example, a plant under American management was manufacturing television sets with a defect rate of 180 defects per 100 sets. The plant was purchased by a Japanese company. After three years of Japanese management the defect rate was reduced to between 3 and 4 defects per 100 sets. The improvement was made with the same work force and equipment. There are many similar stories.

A few American companies have managed to make dramatic improvements under American management. Two of the most noteworthy are Nashua Corporation of New Hampshire and Ford Motor Company. As a consultant to Ford and their suppliers, I have been involved with their program from the beginning. As a consultant to many other companies, I have also had the opportunity to note differences in the approach taken by others who haven't done as well. Ford and Nashua have realized, and the others have not that:

The change requires the *active* leadership of top management (as in the CEO and staff).

The leadership must be obsessed with quality: mere dabbling won't do.

Newer or better techniques are not the answer. Adopting JIT, or TQC, or quality circles, or even all of the techniques will accomplish little without a change in the underlying philosophy and management of the company.

The change must be total. The new philosophy isn't an addition to the current management system; it's a replacement.

The focus of the company must be external, not internal. The company must actively pursue input, including critical input, from customers, suppliers, and employees.

The Ford and Nashua programs are based on Deming's philosophy. Ford's experience is described by Bill Scherkenbach in his excellent book, which is listed in the bibliography (Scherkenbach, 1986).

SUMMARY

This chapter discussed the history of quality in Japan. Deming's 14 points for management were given. The major contributions of the American and Japanese to the quality field were considered. The major differences in the American and Japanese approaches to quality control were examined. A method of measuring or "expressing" quality was shown. Reasons why quality circles so often fail in American firms were provided. Different approaches to vendor and employee relations were offered. The "AQL fallacy" was exposed. Finally, the key ingredients of the successful American programs of quality improvement were provided.

The chapter did not provide a detailed road map for improving quality. In fact, although some maintain that it is possible to provide a generic road map for improving quality, this writer doubts that it is possible. However, the literature does offer a number of success stories from which the interested executive can glean a great deal of useful anecdotal information.

10
The Malcolm Baldrige National Quality Award

HISTORY

In the early 1980s it was apparent that American productivity, while still the highest in the world, was growing at a slower rate than that of many competing foreign countries. It was also obvious that this situation was already having an adverse impact on the American economy and that conditions would get worse if a remedy was not found. The nation's standard of living could not be maintained or improved in the face of declining relative productivity. As a result of this concern, several steps were taken by individuals and groups in both private and public sectors. Some of these steps are described below (DeCario and Sterett, 1990):

In October 1982, President Ronald Reagan signed legislation mandating a nationwide study on productivity.

Much of the information in this chapter is based on materials used in the Malcolm Baldrige Examiner Training Program. The author thanks the National Institute of Standards and Technology and the Malcolm Baldrige National Quality Award Consortium, Inc., for their kind permission to use this material.

In February 1982, the National Advisory Council for Quality (NACQ), a group composed of private and public sector executives was formed. The goal of NACQ was to provide training, publications, and research in the quality disciplines.

The American Productivity and Quality Center (APQC) was formed. The nonprofit organization was concerned with improving quality, productivity, and competitiveness.

A White House Conference on Productivity was sponsored in September 1983. Key speakers included President Reagan, Vice-President George Bush, Treasury Secretary Donald Regan, and Commerce Secretary Malcolm Baldrige.

The report on the White House Conference on Productivity, published in April 1984, called for a national medal to be awarded in recognition of high levels of productivity achieved by American organizations. The close association of productivity and quality became immediately apparent, and quality was soon the focus of attention. In the intervening years, a massive amount of work by various individuals and organizations resulted in the passage of public law 100-007 in August 1987, which established the Malcolm Baldrige National Quality Award, named after the late Commerce Secretary. The first awards were presented by President Reagan in October, 1988.

The scoring criteria has evolved over the years. Table 10.1 summarizes the changes since 1988.

It can be seen that the leadership category lost 50 points and the quality results category gained 50 points. The logic seems to be "the

Table 10.1 Major Categories and Points

Category	1988 Pts	1990 Pts	Change
Leadership	150	100	−50
Information and analysis	75	60	−15
Strategic quality planning	75	90	+15
Human resource utilization	150	150	NC
Quality assurance of products and services	150	150	NC
Quality results	100	150	+50
Customer satisfaction	300	300	NC

Figure 10.1 The Malcolm Baldrige National Quality Award. (Photo used by permission of the National Institute of Standards and Technology.)

proof is in the pudding," i.e., if effective leadership exists, it will be seen in improvements and overall excellence. The other two categories with changed scores are information and analysis, which lost 15 points and strategic quality planning, which gained 15 points. This shifts emphasis from "after the fact" feedback and follow-up to "before the fact" planning.

AWARD CATEGORIES AND NUMBER OF AWARDS

The law provides for up to two awards in each of three separate categories:

1. Manufacturing companies or subsidiaries
2. Service companies or subsidiaries
3. Small businesses

Any for profit business or subsidiary located in the United States may apply. Complete eligibility requirements are given in the Application Guidelines (United States Department of Commerce, 1990).

It is worth noting that there is no requirement to give all of the awards. In the opinion of the examiners and judges, if none of the applicants meet the high standards of the award, it will be withheld. During the first two years, only five of a possible 12 awards were presented; no awards were presented in the service category. This has led to accusations by some that the award criteria has an antiservice bias.

Whether or not the criteria for service businesses will be changed in the future remains to be seen.

OVERVIEW OF THE APPLICATION GUIDELINES

Scoring

The material in this section is based on the 1990 criteria. Since the guidelines change each year, a current copy of the Application Guidelines should be obtained. However, the criteria seem to be stabilizing: compared to the 1988 to 1989 changes, the changes made from 1989 to 1990 were relatively minor. The relationship of the various categories is shown in Figure 10.2.

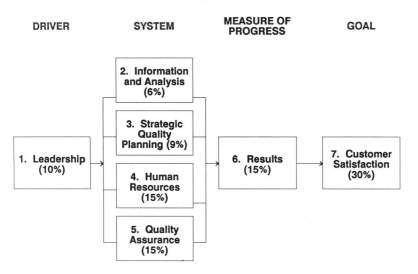

Figure 10.2 Relationship of Baldrige Award categories. (Courtesy of the American Society for Quality Control, reprinted with permission.)

Each of the major categories is further subdivided. The 1990 examination categories, items, descriptions, and point values are as follows:

1.0 Leadership (100 points).
Examines how the senior executives create and sustain a clear and visible quality value system along with a supporting management system to guide all activities of the company toward quality excellence. Also examined are the senior executives' and the company's quality leadership in the external community and how the company integrates its public responsibilities with its quality values and practices.

1.1 Senior Executive Leadership (30 pts.).
Describes the senior executives' leadership, personal involvement, and visibility in developing and maintaining an environment for quality excellence.
1.2 Quality Values (20 pts.).
Describes the company's quality values, how they are projected in a consistent manner, and how adoption of the values throughout the company is assessed and reinforced.

1.3   Management for Quality (30 pts.).
      Describes how the company integrates its quality values into
      day-to-day management of all units.
1.4   Public Responsibility (20 pts.).
      Describes how the company extends its quality leadership to
      the external community and integrates its responsibilites to
      the public for health, safety, environmental protection, and
      ethical business practice into its quality policies and activ-
      ities.

2.0   Information and Analysis (60 pts.).
      Examines the scope, validity, use, and management of data and
      information that underlie the company's total quality manage-
      ment system. Also examined is the adequacy of the data and infor-
      mation to support a responsive prevention approach to quality
      based upon "management by fact."

      2.1   Scope and Management of Quality Data and Information
            (35 pts.).
            Describes the company's base of data and information used
            for planning, management, and evaluation of quality; and
            how data and information reliability, timeliness, and access
            are assured.
      2.2   Analysis of Quality Data and Information (25 pts.).
            Describes how data and information are analyzed to support
            the company's key quality leadership objectives in a time-
            ly manner.

3.0   Strategic Quality Planning (90 pts.)
      Examines the company's planning process for retaining or
      achieving quality leadership and how the company integrates
      quality improvement planning into overall business planning.
      Also examined are the company's short-term and long-term
      priorities to achieve and/or sustain a quality leadership po-
      sition.

      3.1   Strategic Quality Planning Process (40 pts.).
            Describes the company's strategic quality planning process
            for short-term (1–2 years) and long-term (3–5 years or more)
            quality leadership and customer satisfaction.

3.2 Quality Leadership Indicators in Planning (25 pts.).
Describes the company's approach to selecting quality-related competitive comparisons and world-class benchmarks to support strategic quality planning.

3.3 Quality Priorities (25 pts.).
Summarizes the company's principal quality priorities and plans for the short term (1–2 years) and long term (3–5 years or more).

4.0 Human Resource Utilization (150 pts.).
Examines the effectiveness of the company's efforts to develop and realize the full potential of the work force, including management and to maintain an environment conducive to full participation, quality leadership, and personal and organizational growth.

4.1 Human Resource Management (30 pts.).
Describes how the company's human resource plans support its quality leadership objectives; summarizes principal short term (1–2 years) and long term (3–5 years or more) priorities.

4.2 Employee Involvement (40 pts.).
Describes the means available for all employees to contribute effectively to the company's quality objectives; summarizes trends in involvement.

4.3 Quality Education and Training (40 pts.).
Describes how the company decides what quality education and training is needed by employees and how it uses the knowledge and skills acquired; summarizes the types of quality education and training received by employees in all employee categories.

4.4 Employee Recognition and Performance Measurement (20 pts.).
Describes how the company's recognition and performance measurement processes support quality improvement; summarizes trends in recognition.

4.5 Employee Well-Being and Morale (20 pts.).
Describes how the company safeguards the health and safety of employees, ensures comfort and physical protection, and maintains a supportive work environment; summarizes trends in employee well-being and morale.

5.0 Quality Assurance of Products and Services (150 pts.). Examines the systematic approaches used by the company for total quality control of goods and services based primarily upon process design and control, including control of procured materials, parts and services. Also examined is the integration of quality control with continuous quality improvement.

    5.1 Design and Introduction of Quality Products and Services (30 pts.). Describes how new or improved products and services are designed and introduced to meet or exceed customer requirements and how processes are designed to deliver according to the requirements.

    5.2 Process and Quality Control (25 pts.). Describes how the processes which produce the company's products and services are controlled and how the company assures that products and services meet design plans or specifications.

    5.3 Continuous Improvement of Processes, Products, and Services (25 pts.). Describes how products and services are continuously improved through optimization and improvement of processes.

    5.4 Quality Assessment (15 pts.). Describes how the company assesses the quality of products, processes, services, and quality practices.

    5.5 Documentation (10 pts.) Describes documentation and other modes of "knowledge preservation" and transfer to support quality assurance, assessment, and improvement.

    5.6 Quality Assurance, Quality Assessment, and Quality Improvement of Support Services and Business Processes (25 pts.). Desecribes how the quality of support services and business services is assured, assessed, and improved.

    5.7 Quality Assurance, Quality Assessment, and Quality Improvement of Suppliers (20 pts.). Describes how the quality of materials, components, and services furnished by other businesses is assured, assessed, and improved.

6.0 Quality Results (150 pts.). Examines quality levels and quality improvement based upon objective measures derived from analysis of customer requirements and expectations and from analysis of business operations. Also examines current quality levels in relation to those of competing firms.

6.1 Quality of Products and Services (50 pts.). Summarizes trends in quality improvement based upon key product and service quality measures derived from customer needs and expectations.

6.2 Comparison of Quality Results (35 pts.). Compares the company's current quality levels with industry averages, industry leaders, and world leaders, based upon the key product and services quality measures reported in 6.1.

6.3 Business Process, Operational, and Support Service Quality Improvement (35 pts.). Summarizes trends in quality improvement, based upon key measures in business processes, operations, and support services.

6.4 Supplier Quality Improvement (30 pts.). Summarizes trends in improvement in quality of supplies and services furnished by other companies, based upon key measures of product service and quality.

7.0 Customer Satisfaction (300 pts.). Examines the company's knowledge of the customer, overall customer service systems, responsiveness, and its ability to meet requirements and expectations. Also examines current levels and trends in customer satisfaction.

7.1 Knowledge of Customer Requirements and Expectations (50 pts.). Describes how the company determines current and future customer requirements and expectations.

7.2 Customer Relationship Management (30 pts.). Describes how the company provides effective management of its relationships with customers and how it ensures continuous improvement of customer relationship management.

7.3 Customer Service Standards (20 pts.). Describes the company's standards governing the direct contact between employees and customers, and how these standards are set and modified.

7.4 Commitment to Customers (20 pts.). Describes the company's commitments to customers on its explicit and implicit promises underlying the company's products and services.

7.5 Complaint Resolution for Quality Improvement (30 pts.). Describes how the company handles complaints, resolves them, and uses complaint information for quality improvement and prevention of recurrence of problems.

7.6 Customer Satisfaction Determination (50 pts.). Describes the company's methods for determining customer satisfaction, how this information is used in quality improvement, and how methods for determining customer satisfaction are improved.

7.7 Customer Satisfaction Results (50 pts.). Briefly summarizes trends in the company's customer satisfaction and in indicators of adverse customer response.

7.8 Customer Satisfaction Comparison (50 pts.). Compares the company's customer satisfaction results and recognition with those of competitors that provide similar products and services.

THE APPLICATION AND EXAMINATION PROCESSES

The application and examination processes involve several steps. In 1990 the steps were:

1. Submit an eligibility application (February – April). Eligibility determination takes no more than 14 days.
2. If eligible, submit the full application (due by the last Wednesday in April).
3. **First-Stage Review.** Four members of the board of examiners, led by a senior examiner, review the application. The panel of judges review the the results and select those applications to refer for consensus review. Efforts are made to ensure that all applicants receive every reasonable consideration to advance to the next stage. Even applicants that are not referred to consensus review receive a thorough analysis by the board of examiners. However, as much as 400 – 500 examiner hours may be spent on winning applications. Given the limited size of the board of examiners, the first-stage review is necessary to provide the time required.
4. **Consensus review.** Four members of the board of examiners, led by a senior examiner, review the application. The panel of judges review the results and select those applications to refer for site visits.
   [Steps 3 and 4 take place in the May – August period.]
5. **Site visit review.** In the August – September period, on site verification reviews are conducted by at least five members of the board of examiners, led by the senior examiner. The site visit

review team develops a report for the panel of judges. The primary objectives of the site visits are to verify the information provided in the application report and to clarify issues and questions raised during review of the report. Site visit plans are prepared jointly by the applicant and site visit team. Site visit reviews consist primarily of interviews by examiner teams and review of pertinent records and data. Applicants are asked to make introductory and concluding presentations. The site visit teams prepare reports for the panel of judges.

6. **Judges' final review.** The panel of judges review all reports and develop a set of recommendations that are made to the National Institute of Standards Technology (NIST). NIST presents the judges' recommendations to the Commerce Secretary, who makes the final award decisions.

7. **Award presentation.** In October or November, the awards are presented in a special ceremony. The President of the United States made the award presentations in 1988 and 1989 and the event was covered by both the general and business media.

8. **Feedback reports.** At the close of the award cycle, all applicants receive feedback reports. These reports describe strengths and weaknesses and recommend areas for improvement.

The Malcolm Baldrige National Quality Award receives no public funding. All expenses are covered by application and site visit fees, or by the Foundation for the Malcolm Baldrige National Quality Award, a privately funded group. A great deal of volunteer effort is involved in administration of the award, including the time of many of the members of the board of examiners. The basic application fees for 1990 were $2,500 for regular applicants, and $1,000 for small businesses. An eligibility application is required before submitting the full application, a $50 nonrefundable fee is charged for the eligibility application. The $50 fee is deducted from the full fee for those eligible to submit the full application. The remaining application review fee may be paid in two installments. Applicants who qualify for site visits (described below) must pay a fee to cover expenses incurred by the site visit team; small businesses are charged half of the rate of companies in the manufacturing or service categories.

The Malcolm Baldrige National Quality Award is probably the best consulting bargain around. The work done by the examiners and judges is enormous, and those selected to serve on the board of examiners are among the leading quality experts in the country. All members of the board of examiners (which includes the panel of judges) are required to sign conflict of interest statements and nondisclosure agreements, and they must abide by a strict code of ethics.

THE APPLICATION EXPERIENCE

During the first two years of the award, the press made a great deal of fuss about the money and effort expended by businesses to apply for the Malcolm Baldrige National Quality Award. Called the "Oscar of American Business," the award brings instant nationwide recognition and a public relations windfall. The winners make routine use of their victory in their advertising. Justly proud, Xerox Corporation took out full-page advertisements in the *Wall Street Journal* announcing their triumph. *USA Today* gave front page space to the 1989 winners, calling the award "Not Your Average Honor."

But winning was not without its costs. Xerox assigned 17 full-time employees to prepare the application and budgeted $800,000 over and above the payroll expense. Other applicants, who did not win, also invested enormous sums in preparing their submissions. Corning estimates its award team put in 7,000 hours. Ford's application involved 80 people. At least one quality expert was quoted as believing that some companies paid more attention to preparing their application than they did to improving quality.

In contrast to these herculean efforts, Globe Metallurgical, Inc. won the 1988 Malcolm Baldrige National Quality Award in the small business category with an application prepared by a Vice President at home in a single weekend.

In my opinion, the ideal application should be simple with a clear focus on facts. Documentation in the form of easily understood tables and graphs are helpful. Slick, glitzy applications are not necessary. The experience of Globe Metallurgical, Inc. demonstrates this.

WINNING STRATEGIES

One of the fundamental purposes of the Malcolm Baldrige National Quality Award is to promote quality improvement in the United States by publicizing the strategies used by the winners to achieve their success. At the time of this writing, there have been five winners. A synopsis of their winning strategies is presented below, in alphabetical order. NIST sponsors forums conducted by the winners each year to describe their strategies in greater depth. Detailed reports may be acquired from the winning companies through NIST.

Globe Metallurgical, Inc. (1988)

Globe Metallurgical, Inc., is a major producer of silicon metal and ferrosilicon products. In 1985, Globe did a self assessment of their quality system as it related to the requirements of Ford's Q-101 quality standard. Several deficiencies were discovered: the system emphasized detection instead of prevention of quality problems, failed to use SPC, lacked quality planning, and did not involve their employees in the improvement process. Globe addressed these issues by training the entire workforce in SPC, establishing the quality manual and the mechanics of their quality system, and educating and training Globe's suppliers in the necessity of SPC and a quality system. Outside consultants were used for SPC training and supervisors were trained in Deming's approach with a videotape program.

Management formed an executive steering committee and plant committees. The executive committee, chaired by the President and CEO, meets monthly to discuss broad issues of development and maintenance of the quality system. Plant committees, chaired by the plant managers, meet daily to discuss specific projects and implementation issues.

Milliken & Company (1989)

Headquartered in Spartanburg, South Carolina, the privately owned Milliken & Company has 14,300 "associates" employed primarily at 47 manufacturing facilities in the United States. Annual sales exceed $1 billion for textile and chemical products. In 1981, senior management implemented Milliken's pursuit of excellence (POE), a commitment to customer satisfaction that pervades all company levels. Since then, productivity has increased 42%.

Teams are a hallmark of the Milliken quality improvement process. In 1988, 1,600 corrective action teams formed to address specific manufacturing or other internal business problems, 200 supplier action teams worked to improve Milliken's relationships with its suppliers, and nearly 500 other teams responded to the needs and aims of the customers. Quality improvement measures are based solidly on factual information, contained in an array of standardized data bases accessible from all Milliken facilities. Most manufacturing processes are under the scrutiny of real-time monitoring systems that detect errors and help pinpoint their causes. Milliken's successful push for quality improvement has allowed it to increase sales in the United States and to enter foreign markets.

Motorola (1988)

At Motorola, the quest for quality involves every aspect of work: from SPC to product development cycles to the time it takes for the order to be entered. In 1987 Motorola invested $44 million in training, much of that strictly on quality improvement. A key factor in Motorola's achievements is participative management process that emphasizes employee involvement at all levels.

In 1981, Motorola launched an ambitious drive for a tenfold improvement in the quality of products and services by 1986. Having largely accomplished that goal in 1986, the company then set its sights even higher, striving for a hundredfold improvement by 1991. The company's next goal is Six Sigma (6σ) quality by 1992.

While the company expresses this goal in the language of statistics (6σ means no more than 3.4 defects per million parts produced), the ultimate goal is zero defects in everything and Motorola's overriding objective is total customer satisfaction. The strategy is to refocus all elements of Motorola's business on serving the customer.

Westinghouse Commercial Nuclear Fuel Division (1988)

Westinghouse Commercial Nuclear Fuels Division (WCNFD) is a division within the Energy Systems Business Unit of Westinghouse. It is responsible for the engineering, manufacturing, and supply of pressurized water reactor (PWR) fuel assemblies for commercial nuclear power reactors. The fuel contained in these assemblies generates heat (through nuclear fission), which is converted to electricity. WCNFD supplies 40% of the United States light water reactor fuel market and 20% of the free world market.

Quality standards are demanding. The reliability of Westinghouse nuclear fuel approaches 99.9995%. The quality challenge is compounded by the complexity of the fuel production process, which involves both chemical and mechanical operations and literally thousands of variables. By 1983, WCNFD recognized that quality could provide the most significant competitive edge. Consequently, the division made continuous, long-term quality improvement its number one management objective. The division was quick to identify and implement new and innovative management techniques and quality improvement processes, including human resource management concepts (participative management and quality teams), new technologies (robotics and artificial intelligence), statistical in process quality control, and advanced management concepts such as cycle-time reduction.

Each year, WCNFD increased its understanding of the quality management process and sharpened the focus and direction of its program. The plan focused on a different theme each year: 1983 increased general employee awareness of the importance of quality to its business; 1984 began tracking and reporting quality failure costs; 1985 established division-wide quality goals and a measurement system to measure progress; 1986 created a division-wide mission statement and added second- and third-tier quality measures; 1987 shifted primary focus to the customer; and 1988 centered on the "Total Quality" theme.

From its experience in quality improvement, WCNFD learned a number of lessons, which it believes can be applied with reasonable success to any business. First, an organization must have a common vision, and all people in that organization must embrace a common mission. Next, a common framework for total quality is absolutely critical. This is the model or blueprint that keeps everyone focused on continuous improvement in all aspects of the business. The third lesson is to measure, measure, measure. WCNFD discovered that no quality improvement program can be successful unless both employees and customers are intimately involved in the process. Finally, total quality requires a long term commitment to continuous quality improvement. A total quality culture cannot be built overnight. Total quality is not a short term proposition, but neither are the rewards; the benefits of total quality are long term for a company willing to make the commitment.

Xerox Business Products and Systems (1989)

In 1983, Xerox Business Products and Systems launched an ambitious quality improvement program to arrest its decline in a world market it had once dominated. Today, the company can reclaim the title of the industry's best in nearly all copier product markets. The company, headquartered in Stamford, Connecticut, attributes the turnaround to its strategy of "leadership through quality." Through extensive data collection efforts, Xerox Business Products and Systems knows what customers want in products and services. Planning of new products and services is based on detailed analyses of data organized in some 375 information management systems, of which 175 are specific to planning, managing, and evaluating quality improvement.

Benchmarking is highly developed at Xerox Business Products and Systems. In all key areas of product, service, and business performance, the company measures its achievement for each attribute and compares itself with the level of performance achieved by the world leader, regardless of industry.

Quality improvement and, ultimately, customer satisfaction are the job of every employee. Working with the union, the company ensures that workers are vested with considerable authority over day-to-day work decisions. Employees are expected to take the initiative in identifying and correcting problems that affect the quality of products or services.

Xerox Business Products and Systems employs 50,200 people at 83 locations in the United States. The company makes more than 250 types of document processing equipment. United States sales exceeded $6 billion in 1988.

BALDRIGE AWARD VERSUS DEMING PRIZE

The Deming Prize, Japanese industry's most coveted prize, was one of the sources of inspiration for the Malcolm Baldrige National Quality Award. Established in 1951, the Deming Prize was named for Dr. Edwards Deming, the celebrated American quality and statistical expert who helped the Japanese overcome their worldwide reputation for poor quality.

There are both similarities and differences between the two

awards. Both programs are intended to raise the quality levels of industries in their respective countries. Both involve applications by interested firms and both have on-site visits by teams of examiners.

Although both awards can be won by large or small businesses in either service or manufacturing sectors, the Deming Prize, unlike the Baldrige Award, can also be won by individuals or individual factories. There is no separate Deming Prize for Service businesses.

Perhaps the key area of difference is that the Baldrige Award puts more emphasis on results while the Deming Prize stresses the process that produces the results. The Deming criteria are very specific in describing quality improvement methods, while the Baldrige criteria reveal virtually nothing on the subject. As would be expected from its results orientation, the Baldrige criteria also place more emphasis on customer satisfaction.

The examination process differs considerably between the two awards. Examiners for the Baldrige Award are carefully selected from experts in academia and industry by a panel of judges each year, determined from a pool of applicants. A certain degree of turnover is built into the examiner selection process. The Deming Prize is administered by the Union of Japanese Scientists and Engineers (JUSE), who evaluate candidate firms with "counselors" who may also advise the firms in a quasi-consultant/client relationship. The American Society for Quality Control and other professional organizations play a very active role in the Baldrige Award, but a role fundamentally different from that of JUSE and the Deming Prize. American professional organizations serve primarily as administrators who provide assistance to NIST.

The Japanese system judges the applicants against predetermined criteria, essentially a qualifying criteria. The applicants for the Deming Prize essentially compete against the criteria, rather than against one another. The Baldrige Award involves a certain degree of qualification by requiring "world-class performance." Qualified applicants are matched against other applicants to compete for the award. Also, Baldrige Award winners must demonstrate to the judges that they provide an appropriate model for other American businesses to follow.

Finally, the Baldrige Award has a built-in continuous improvement mechanism. A board of overseers evaluates feedback from applicants, examiners, judges, and other sources and recommends modifications to the program if deemed necessary.

THE FUTURE OF THE BALDRIGE AWARD

The Malcolm Baldrige National Quality Award is not without its critics. These critics say:

Its criteria has an antiservice bias.

It encourages applicants to win, but diverts attention from quality improvement.

It takes too much time, money, and effort to apply.

The eventual flood of applications will water down the award by forcing the use of less qualified examiners.

Its emphasis on results encourages the same short-term "bottom line mentality" that is partially responsible for America's loss of competitiveness in the first place.

The examination process is so difficult and so time-consuming that examiners will become burned out.

In my opinion there is some validity in these criticisms. As an examiner in 1988, the first year of the award, I was among the critics of the Baldrige Award. However, nearly every member of board of examiners was also among the award's critics; criticism is actively solicited by those responsible for the award. The board of overseers responded by changing the award criteria, category weights, and even the examination process itself. In the first three years of existence, the award criteria have changed considerably. Continuity has been maintained by keeping the seven basic categories intact.

In short, although the award is far from perfect, its built-in improvement mechanism is already working to make it better. In my opinion, the award is a positive force in the effort to improve quality in America. I believe the award will improve steadily, gain recognition, and its influence will grow.

11
Total Quality Management

Total Quality Management (TQM) comes in a variety of "flavors." This chapter describes the version of TQM proposed by the Department of Defense (DoD) for their internal operations and suppliers, and is described in the publication DoD 5000.51-G, "Total Quality Management Guide." Table 11.1 presents a draft of the DoD's TQM directive (received by the author from the DoD on April 4, 1990).

TQM presents little in the way of revolutionary ideas. It is, essentially, an implementation of the principles and methodology descibed throughout this book. This does not in any way diminish the value of the DoD's TQM initiative. Indeed, the endorsement of the principles and methodology by DoD is likely to result in the acceleration of their acceptance by defense contractors.

There is some doubt as to the ability of DoD, or anyone else, to *mandate* TQM. As one executive involved in promoting TQM said (Hayes, 1990), "The change to a quality-driven division is not managed or mandated; it is led." The very essence of TQM is a philosophy of excellence. If the philosophy is mere grandstanding to appease DoD

195

Table 11.1   Draft DoD Directive on TQM

SUBJECT: Total Quality Management

References:
a. Executive Order 12637, "Productivity Improvement Program for the Federal Government," April 27, 1988.
b. DoD 5000.51-G, "Total Quality Management Guide," authorized by this Directive.
c. DoD 5025.1-M, "Directives System Procedures," April 1981, authorized by DoD Directive 5025.1, December 23,1988.

A. PURPOSE

This Directive:
1. Establishes policy and assigns responsibilities under reference a. for the implementation of the Total Quality Management (TQM) concept in the Department of Defense. TQM is the vehicle to drive out waste and maximze effectiveness, innovation, productivity, quality of work life, and providing products and services that satisfy or exceed customer requirements at a cost that represents best value.
2. Authorizes publication of DoD 5000.51-G, "Total Quality Management Guide" b., in accordance with reference c.

B. APPLICABILITY

This Directive applies to the Office of the Secretary of Defense (OSD), the Military Departments, the Joint Chiefs of Staff (JCS), and the Defense Agencies.

C. DEFINITION

Total Quality Management (TQM). A philosophy and a set of guiding principles that represent the foundation of a continuously improving organization. It is the application of quantitative methods and human resources to improve the material and services supplied to an organization, all the processes within an organization, and the degree to which the needs of the customer are met; now, and in the furture. It integrates fundamental management techniques, existing improvement efforts, and technical tools under a disciplined approach focused on continuous improvement.

Table 11.1 *(continued)*

D. POLICY

It is DoD policy that:

1. Principles of TQM must involve all DoD personnel, processes, products, and services, including the generation of products in paper and data form.
2. Process management, process improvement, and process measurement are fundamental management approaches that are to be used, as appropriate, by all DoD managers.
3. TQM concepts are to be among the fundamental management tenets of every DoD activity and are to be ingrained throughout the Department of Defense with tailored training for each organizational level, starting with top management.
4. Managers and personnel at all levels must take responsibility for the quality of their processes and products. Accurate quantitative measures of quality should be established as a basis for informed improvement action.
5. Involved, competent, and, dedicated employees make the greatest contributions to quality and productivity. They must be recognized and rewarded accordingly.
6. Acquisition strategies will address plans to measure and pursue continuous process improvement to provide products and services that will provide best value.
7. TQM will be a key consideration in source selection.
8. Emphasis must change from relying on inspection to designing and building quality into the processes that affect product quality.
9. Technology (being one of our greatest assets) must be used, where appropriate, to continuously improve the quality of defense systems, equipment, and services.
10. Continuous process improvement is a key to performance improvement and must be pursued with the necessary resources to achieve the desired cultural change in the Department of Defense. (See DoD 5000.51-G, ref. b.)

E. RESPONSIBILITIES

1. The Defense Council on Integrity and Management Improvement (DCIMI), under the leadership of the Secretary of Defense and the Deputy Secretary of Defense, will function as the Executive Steering Committee for TQM.
2. The Under Secretary of Defense (Acquisition) (USD(A)) will:
    a. Act as the OSD office of primary responsibility for development of TQM policy and procedures.

Table 11.1   *(continued)*

    b.   Ensure commonality of TQM training and implementation throughout the Department of Defense.

3.   The Secretaries of the Military Departments, the Joint Chiefs of Staff (JCS), and the Directors of Defense Agencies will:

    a.   Implement DoD policy on TQM.

    b.   Provide the leadership and management necessary to implement TQM in their organizations.

    c.   Develop and maintain a TQM implementation plan.

    d.   Provide appropriate training in TQM principles and techniques.

F. EFFECTIVE DATE AND IMPLEMENTATION

This Directive is effective immediately. DoD components will forward one copy of implementing documents to the Under Secretary of Defense (Acquisition) within 120 days.

---

auditors it will fail to provide the necessary foundation for TQM. Many DoD contractors believe that the government requires that its suppliers use TQM (Hayes, 1990). However, DoD makes no pretense of requiring TQM as a contractural obligation. In fact, DoD documents indicate that Directive 5000.43, "Acquisition Streamlining," probably precludes making TQM a contractural mandate. However, DoD has made it very clear that TQM will be a consideration in awarding future contracts. Given this fact, and economic realities, the wise bidder will take a serious look at TQM.

GOALS

In August 1988, DoD set out three sets of goals in their TQM Master Plan: short-range (1989), mid-range (1991), and long-range (1995). The goals are given here for two reasons. First, they are of interest in and of themselves; second, they are instructional for DoD contractors and others who are interested in implementing TQM.

Short-Range Goals (1 year)

1.   Establish executive steering committee and subordinate teams, and begin training and continuous process improvement activities.

2. Identify initial cadre of TQM facilitators, and begin facilitating team activities and training staff facilitators.
3. Implement the TQM training strategy and begin collecting and developing training materials.
4. Establish an R & D program in support of TQM. (This program will capture and disseminate information on the DoD implementation of TQM).
5. Develop and implement recognition and reward systems based on TQM goals and behaviors.
6. Begin ensuring consistency among TQM and major documentation and guidance.
7. Begin enlisting defense industry commitment.

Mid-Range Goals (3 years)

1. Establish and implement policy deployment mechanisms.
2. Harmonize DoD directives/regulations/instructions and TQM.
3. Eliminate barriers to TQM implementation.
4. Implementation commitment by major defense contractors, with "critical mass" achieved in at least the top 25 contractors. ("Critical mass" is defined as the senior leadership being trained and actively performing continuous process improvement. As well, selected programs or departments should have begun full process improvement implementation.)
5. DoD acquisition personnel use TQM principles and practices in dealing with industry.
6. Develop, produce, acquire, and promulgate a standard set of TQM training materials.
7. Establish a mature, functioning staff of facilitators.
8. Understand and coordinate with TQM efforts by other sectors of the federal government.
9. Develop and cultivate key congressional TQM champions.

Long-Range Goals (7 years)

1. Establish TQM as a way of life.
2. All DoD personnel directly doing continuous process improvement.

3. Widespread defense industry implementation of continuous process improvement.
4. Congressional understanding of and support for TQM.*

TQM EDUCATION AND TRAINING

The DoD has four long-range goals relating to TQM training (DoD, July 1989).

1. All DoD personnel will be educated and trained in the principles of TQM and in the tools needed to implement it.
2. TQM will be integrated into all appropriate DoD management and technical education and training courses.
3. DoD will have a support network of resources and communications internal and external to DoD.
4. DoD will encourage private and public educational systems and professional associations to provide and foster TQM education.

DoD does not explicitly define the meaning of long-range, so the time frame involved is only a guess. However, since training is a necessary precondition for the overall goals of TQM described above, it seems reasonable to assume that DoD plans to reach these goals in something less than 5 years (by 1995).

Since management is the key to TQM, management training is of obvious importance. Two management awareness courses are proposed by DoD. Senior Management Awareness is a two-day course with the objective of providing participants with an understanding of the basic concepts of TQM. Participants will also understand management's role and expected behaviors. Table 11.2 presents the curriculum for the proposed course.

In addition to the Senior Management Awareness course, DoD proposes a second course for general management, which would last eight days and would provide exposure to quality awareness and TQM philosophy; quantitative methods to analyze processes; group development skills, essential in building trust and sharing ideas within the

---

*As of this writing, DoD has a long way to go in this area; lawmakers deleted TQM implementation funds from the defense budget in the summer of 1989.

Table 11.2 Curriculum for Two-Day TQM Management Awareness Course

A. What is meant by TQM?
  1. TQM history and background
  2. Definitions and general explanation of basic principles:
     a. TQM
     b. Quality
     c. Process
     d. Variation
     e. Continuous Improvement
B. Management's role and behaviors during TQM implementation:
  1. Committed leader (public expression of intent and actions they will take)
  2. Active, involved participant
  3. Change agent of organizational culture through effective communication and teamwork
  4. Provider of resource support
C. Basic methods and tools of analysis:
  1. Plan-Do-Check-Act (PDCA) cycle
  2. Basic statistical tools

workplace; and knowledge about how to change organizational culture and overcome obstacles to implementing TQM.

This workshop is organized into two sessions, one lasting five days and the other three days. The two sessions are seperated by a three week "back at work" interval. The first five-day session is devoted to quality awareness, systems thinking, statistical thinking, the concept of process improvement, decision making, and interpersonal skills. Upon return, the participants will be involved in training in the additional statistical tools that aid in summarizing the data collected at the workplace, learning about additional TQM concepts, and learning how to recognize and overcome barriers they are likely to encounter. Table 11.3 outlines the curriculum for the eight-day course.

The DoD envisions that education will be provided from a variety of sources. Specifically, work is being done to develop "delivery agents" internal to DoD, in federal government agencies other than DoD, from state and local government academic institutions, and from the private sector. It is possible that the demand will create its own supply. However, DoD is lagging the private sector by about a decade in im-

Table 11.3   Curriculum for Eight-Day TQM General Management Awareness Course

A. Overview of the quality problem:
  1. The real threat and challenge of foreign competition
  2. Quality of American products and services
  3. Examples of DoD quality problems
B. History of quality and quality management worldwide and in the United States:
  1. Post World-War II management philosophies and practices
  2. management and worker's attitudes and relationships
  3. History of TQM in DoD
C. TQM guiding principles concerning relationship of quality and productivity and cost:
  1. Quality-productivity-cost relationship (chain reaction of quality and productivity)
  2. Detection versus prevention
  3. The cost of quality
D. TQM key concepts:
  1. Production viewed as a system
  2. The paradigm shift
  3. Continuous improvement
  4. Management of processes versus management of outcomes
  5. Variation: common causes and special causes
  6. Customer orientation
  7. Cross-functional management teams
E. Strategies for process improvement (Plan-Do-Check-Act cycle)
F. Responsibilities of management in TQM implementation:
  1. Team building and group facilitation
  2. Communication (breaking down barriers, developing and supporting effective horizontal and vertical communication)
  3. Leadership (should be "follow me" approach; management must lead by example, be mentors, guides, and supporters)
  4. Transforming organizational culture
G. Understanding graphic tools:
  1. Application of tools and procedures
    a. Flow chart
    b. Cause-and-effect (fishbone) diagram
    c. Pareto chart
    d. Affinity diagram
    e. Histogram
    f. Scatter diagram
    g. Check sheet

    h. Run chart
    i. Control Chart
    j. Matrix diagram
    k. Tree diagram
    l. Process decision program chart
    m. Arrow diagram
  2. Appropriateness of tools for particular situations
H. Introduction to data collection:
  1. Collection of meaningful data
  2. Qualitative and quantitative data
I. Basic statistics:
  1. Statistical theory
  2. Statistical thinking
  3. Statistical approach to continuous improvement
J. Measurement:
  1. Factors that affect measurement
  2. Outcome versus process measures

---

plementing continuous quality improvement. Thus, a good deal of the required infrastructure already exists. It remains to be seen if the additional demand created by defense contractors can be met. The most difficult resource to acquire, and by far the most important, will be qualified instructors. Qualified individuals will need an ability to teach, extensive credentials in theoretical and applied statistics, and broad experience in industry. This mix of skill is by no means common.

DoD has proposed establishment of a TQM resource center to disseminate information. The resource center would serve as a clearinghouse, with computer search facilities and experienced consultants to help training developers select existing training modules to use in their own programs. It would also provide support during all phases of implementation. A case study data bank would be developed to disseminate knowledge acquired from TQM implementation. An on-line TQM applications journal would be created and updated either monthly or quarterly.

TQM KEY FEATURES

TQM is an implementation of existing methods of quality improvement in the context of an organizational environment dedicated to con-

Table 11.4 Common Tools Used in Implementing TQM

Tools & Techniques (Listed without regard to priority)	Problem-Solving Activities					
	Bounds & Prioritize Problems	Compile Information	Analysis	General Alternatives	Evaluate	Plan & Implement
Benchmarking	✓	✓	✓		✓	
Cause & Effect Diagrams	✓		✓			
Nominal Group Technique	✓	✓		✓	✓	✓
Quality Function Deployment	✓		✓		✓	
Pareto Charts	✓	✓	✓			
Statistical Process Control		✓	✓			
Histograms		✓	✓			
Check Sheets		✓	✓			
Input/Output Analysis			✓			
Scatter Diagrams			✓			
Concurrent Engineering			✓	✓	✓	
Design of Experiments			✓	✓	✓	
Cost of Quality			✓		✓	
Control Charts			✓		✓	
Work Flow Analysis			✓			✓
Team Building	✓					✓
Time Management	✓		✓	✓	✓	✓
Shewhart Cycle		✓	✓			✓

tinuous improvement. Table 11.4 presents the TQM tools most commonly used and where they are usually applied (DoD, Vol. 1, Feb. 1990).

Most of these tools and techniques are described elsewhere in this book. The recommended reading at the end of the book provides additional background.

STARTING TQM

TQM is an activity that is "people intensive." It requires only a small amount of investment in hardware. Since time is a scarce resource it is helpful to have a strategy for using it effectively. The "General Strategy for Getting Started" recommended by DoD is summarized below (DoD, Vol. 2, Feb. 1990).

Demonstrate Leadership

1. Take the initiative
2. Demonstrate commitment
3. Create more leaders
4. Guide the efforts of others
5. Remove roadblocks and barriers

Build Awareness

1. Build your personal awareness
2. Read the suggested literature and view the suggested videotapes
3. Attend TQM-related symposia, workshops, and conferences
4. Help your manager become more aware
5. Discuss TQM with your peers
6. Help subordinates become more aware
7. Share the concept with union leadership
8. Inform customers and suppliers

Open and Maintain Lines of Communication

1. Open the communication processes
2. Keep everyone informed

Create a Constancy of Purpose

1. Examine your organization's mission
2. Take a long-term view
3. Establish meaningful goals
4. Effectively deploy policy throughout the organization
5. Treat goals and performance carefully
6. Ensure that behavior and signals are consistent with goals
7. Align overall improvement activity with organizational goals

Focus on the Customer

1. Link organizational purpose to customer satisfaction
2. Identify external and internal customers
3. Understand customer needs, expectations, and requirements
4. Establish routine and meaningful dialogue with customers
5. Listen to the customer
6. Involve the customer in planning and decision making

Choose Early Efforts in Visible Areas Critical to Success

1. Address critical issues that are also important to customers
2. Start with the processes you own and work to improve the system
3. Start the process at the highest level possible
4. Cascade the process through the organization

Develop Teamwork

1. Facilitate team development and activity
2. Guide teams in employing a systematic approach to improvement
3. Create cross-functional teams for cross-functional purposes
4. Create special teams for special issues
5. Work toward everyone becoming part of an improvement team
6. Support and reinforce team behavior and performance
7. Recognize effort and approach, not just results
8. Ensure that process-improvement teams are linked vertically and horizontally

Provide Support, Training, and Education

1. Make improvement a high-priority activity
2. Create time for your people to address improvement
3. Organize to support improvement
4. Educate everyone consistently
5. Make full use of available resources
6. Provide training and education resources
7. Train the managers and supervisors

Build Trust and Respect

1. Recognize that people are your most important treasure
2. Value and encourage individual contributions
3. Recognize that everyone has a responsibility for your quality
4. Listen for even the smallest voice

Create an Environment in Which Continuous Improvement is a Way of Life

1. Expect improvement
2. View problems as opportunities
3. Constantly examine the value of policies, practices, and procedures
4. Drive out the source of fear
5. Recognize success and share the credit
6. Assess improvements based on data, not intuition
7. Encourage innovative thinking and new ideas
8. Align the reward and recognition systems with the TQM philosophy
9. Cease reliance on mass inspection
10. Employ meaningful measurement systems
11. Cost-justify intelligently

Continuously Improve all Processes

1. Understand that everything involves a process
2. Develop process ownership

3. Ensure that every team owns its process
4. Carefully define your processes
5. Study process variation
6. Bring processes under control
7. Standardize and improve on current best-known methods
8. Assess process capability
9. Remove complexity: simplify, consolidate, eliminate
10. Encourage small incremental improvements

Expand Culture to Suppliers

1. Simplify the acquisition process
2. Involve suppliers early
3. Engage suppliers in mutual problem-solving
4. Help suppliers improve
5. Reward improving suppliers appropriately
6. If possible, minimize the number of suppliers
7. Go with the few best and improvement-oriented suppliers
8. Listen to suppliers
9. Remove roadblocks and barriers instead of creating them
10. Work to build mutual trust and respect

Conclusion

This chapter described the DoD's implementation of Total Quality Management. TQM is essentially a focus on providing the customer with continuously improving quality. Implementation begins with the strong, active leadership from the top. It uses SPC and other methods that are well known and proven. Although DoD does not require TQM as a contractural obligation, it does weigh TQM heavily in the awarding of contracts.

Appendix

Table 1   Cumulative Normal Distribution--Values of P.

Values of P corresponding to z_p for the normal curve.

z is the standard normal variable.   The value of P for $-z_p$ equals one minus the value of P for $+z_p$,
e.g., the P for -1.62 equals $1 - .9474 = .0526$.

z_p	.00	.01	.02	.03	.04	.05	.06	.07	.08	.09
.0	.5000	.5040	.5080	.5120	.5160	.5199	.5239	.5279	.5319	.5359
.1	.5398	.5438	.5478	.5517	.5557	.5596	.5636	.5675	.5714	.5753
.2	.5793	.5832	.5871	.5910	.5948	.5987	.6026	.6064	.6103	.6141
.3	.6179	.6217	.6255	.6293	.6331	.6368	.6406	.6443	.6480	.6517
.4	.6554	.6591	.6628	.6664	.6700	.6736	.6772	.6808	.6844	.6879
.5	.6915	.6950	.6985	.7019	.7054	.7088	.7123	.7157	.7190	.7224
.6	.7257	.7291	.7324	.7357	.7389	.7422	.7454	.7486	.7517	.7549
.7	.7580	.7611	.7642	.7673	.7704	.7734	.7764	.7794	.7823	.7852
.8	.7881	.7910	.7939	.7967	.7995	.8023	.8051	.8078	.8106	.8133
.9	.8159	.8186	.8212	.8238	.8264	.8289	.8315	.8340	.8365	.8389
1.0	.8413	.8438	.8461	.8485	.8508	.8531	.8554	.8577	.8599	.8621
1.1	.8643	.8665	.8686	.8708	.8729	.8749	.8770	.8790	.8810	.8830
1.2	.8849	.8869	.8888	.8907	.8925	.8944	.8962	.8980	.8997	.9015
1.3	.9032	.9049	.9066	.9082	.9099	.9115	.9131	.9147	.9162	.9177
1.4	.9192	.9207	.9222	.9236	.9251	.9265	.9279	.9292	.9306	.9319
1.5	.9332	.9345	.9357	.9370	.9382	.9394	.9406	.9418	.9429	.9441
1.6	.9452	.9463	.9474	.9484	.9495	.9505	.9515	.9525	.9535	.9545
1.7	.9554	.9564	.9573	.9582	.9591	.9599	.9608	.9616	.9625	.9633
1.8	.9641	.9649	.9656	.9664	.9671	.9678	.9686	.9693	.9699	.9706
1.9	.9713	.9719	.9726	.9732	.9738	.9744	.9750	.9756	.9761	.9767
2.0	.9772	.9778	.9783	.9788	.9793	.9798	.9803	.9808	.9812	.9817
2.1	.9821	.9826	.9830	.9834	.9838	.9842	.9846	.9850	.9854	.9857
2.2	.9861	.9864	.9868	.9871	.9875	.9878	.9881	.9884	.9887	.9890
2.3	.9893	.9896	.9898	.9901	.9904	.9906	.9909	.9911	.9913	.9916
2.4	.9918	.9920	.9922	.9925	.9927	.9929	.9931	.9932	.9934	.9936
2.5	.9938	.9940	.9941	.9943	.9945	.9946	.9948	.9949	.9951	.9952
2.6	.9953	.9955	.9956	.9957	.9959	.9960	.9961	.9962	.9963	.9964
2.7	.9965	.9966	.9967	.9968	.9969	.9970	.9971	.9972	.9973	.9974
2.8	.9974	.9975	.9976	.9977	.9977	.9978	.9979	.9979	.9980	.9981
2.9	.9981	.9982	.9982	.9983	.9984	.9984	.9985	.9985	.9986	.9986
3.0	.9987	.9987	.9987	.9988	.9988	.9989	.9989	.9989	.9990	.9990
3.1	.9990	.9991	.9991	.9991	.9992	.9992	.9992	.9992	.9993	.9993
3.2	.9993	.9993	.9994	.9994	.9994	.9994	.9994	.9995	.9995	.9995
3.3	.9995	.9995	.9995	.9996	.9996	.9996	.9996	.9996	.9996	.9997
3.4	.9997	.9997	.9997	.9997	.9997	.9997	.9997	.9997	.9997	.9998

Table 2 Poisson Distribution. Probabilities of c or less, given c', appears in the body of the table multiplied by 1000.

c' or np' \ c	0	1	2	3	4	5	6	7	8	9
0.02	980	1,000								
0.04	961	999	1,000							
0.06	942	998	1,000							
0.08	923	997	1,000							
0.10	905	995	1,000							
0.15	861	990	999	1,000						
0.20	819	982	999	1,000						
0.25	779	974	998	1,000						
0.30	741	963	996	1,000						
0.35	705	951	994	1,000						
0.40	670	938	992	999	1,000					
0.45	638	925	989	999	1,000					
0.50	607	910	986	998	1,000					
0.55	577	894	982	998	1,000					
0.60	549	878	977	997	1,000					
0.65	522	861	972	996	999	1,000				
0.70	497	844	966	994	999	1,000				
0.75	472	827	959	993	999	1,000				
0.80	449	809	953	991	999	1,000				
0.85	427	791	945	989	998	1,000				
0.90	407	772	937	987	998	1,000				
0.95	387	754	929	984	997	1,000				
1.00	368	736	920	981	996	999	1,000			
1.1	333	699	900	974	995	999	1,000			
1.2	301	663	879	966	992	998	1,000			
1.3	273	627	857	957	989	998	1,000			
1.4	247	592	833	946	986	997	999	1,000		
1.5	223	558	809	934	981	996	999	1,000		
1.6	202	525	783	921	976	994	999	1,000		
1.7	183	493	757	907	970	992	998	1,000		
1.8	165	463	731	891	964	990	997	999	1,000	
1.9	150	434	704	.875	956	987	997	999	1,000	
2.0	135	406	677	857	947	983	995	999	1,000	

Reproduced with permission from I. W. Burr, *Engineering Statistics and Quality Control*, McGraw-Hill, New York, 1953, pp. 417-421.

c' or np' \ c	0	1	2	3	4	5	6	7	8	9
2.2	111	355	623	819	928	975	993	998	1,000	
2.4	091	308	570	779	904	964	988	997	999	1,000
2.6	074	267	518	736	877	951	983	995	999	1,000
2.8	061	231	469	692	848	935	976	992	998	999
3.0	050	199	423	647	815	916	966	988	996	999
3.2	041	171	380	603	781	895	955	983	994	998
3.4	033	147	340	558	744	871	942	977	992	997
3.6	027	126	303	515	706	844	927	969	988	996
3.8	022	107	269	473	668	816	909	960	984	994
4.0	018	092	238	433	629	785	889	949	979	992
4.2	015	078	210	395	590	753	867	936	972	989
4.4	012	066	185	359	551	720	844	921	964	985
4.6	010	056	163	326	513	686	818	905	955	980
4.8	008	048	143	294	476	651	791	887	944	975
5.0	007	040	125	265	440	616	762	867	932	968
5.2	006	034	109	238	406	581	732	845	918	960
5.4	005	029	095	213	373	546	702	822	903	951
5.6	004	024	082	191	342	512	670	797	886	941
5.8	003	021	072	170	313	478	638	771	867	929
6.0	002	017	062	151	285	446	606	744	847	916

	10	11	12	13	14	15	16
2.8	1,000						
3.0	1,000						
3.2	1,000						
3.4	999	1,000					
3.6	999	1,000					
3.8	998	999	1,000				
4.0	997	999	1,000				
4.2	996	999	1,000				
4.4	994	998	999	1,000			
4.6	992	997	999	1,000			
4.8	990	996	999	1,000			
5.0	986	995	998	999	1,000		
5.2	982	993	997	999	1,000		
5.4	977	990	996	999	1,000		
5.6	972	988	995	998	999	1,000	
5.8	965	984	993	997	999	1,000	
6.0	957	980	991	996	999	999	1,000

Table 2 (continued).

c' or np'	0	1	2	3	4	5	6	7	8	9
6.2	002	015	054	134	259	414	574	716	826	902
6.4	002	012	046	119	235	384	542	687	803	886
6.6	001	010	040	105	213	355	511	658	780	869
6.8	001	009	034	093	192	327	480	628	755	850
7.0	001	007	030	082	173	301	450	599	729	830
7.2	001	006	025	072	156	276	420	569	703	810
7.4	001	005	022	063	140	253	392	539	676	788
7.6	001	004	019	055	125	231	365	510	648	765
7.8	000	004	016	048	112	210	338	481	620	741
8.0	000	003	014	042	100	191	313	453	593	717
8.5	000	002	009	030	074	150	256	386	523	653
9.0	000	001	006	021	055	116	207	324	456	587
9.5	000	001	004	015	040	089	165	269	392	522
10.0	000	000	003	010	029	067	130	220	333	458

	10	11	12	13	14	15	16	17	18	19
6.2	949	975	989	995	998	999	1,000			
6.4	939	969	986	994	997	999	1,000			
6.6	927	963	982	992	997	999	999	1,000		
6.8	915	955	978	990	996	998	999	1,000		
7.0	901	947	973	987	994	998	999	1,000		
7.2	887	937	967	984	993	997	999	999	1,000	
7.4	871	926	961	980	991	996	998	999	1,000	
7.6	854	915	954	976	989	995	998	999	1,000	
7.8	835	902	945	971	986	993	997	999	1,000	
8.0	816	888	936	966	983	992	996	998	999	1,000
8.5	763	849	909	949	973	986	993	997	999	999
9.0	706	803	876	926	959	978	989	995	998	999
9.5	645	752	836	898	940	967	982	991	996	998
10.0	583	697	792	864	917	951	973	986	993	997

	20	21	22
8.5	1,000		
9.0	1,000		
9.5	999	1,000	
10.0	998	999	1,000

c / c′ or np′	0	1	2	3	4	5	6	7	8	9
10.5	000	000	002	007	021	050	102	179	279	397
11.0	000	000	001	005	015	038	079	143	232	341
11.5	000	000	001	003	011	028	060	114	191	289
12.0	000	000	001	002	008	020	046	090	155	242
12.5	000	000	000	002	005	015	035	070	125	201
13.0	000	000	000	001	004	011	026	054	100	166
13.5	000	000	000	001	003	008	019	041	079	135
14.0	000	000	000	000	002	006	014	032	062	109
14.5	000	000	000	000	001	004	010	024	048	088
15.0	000	000	000	000	001	003	008	018	037	070

	10	11	12	13	14	15	16	17	18	19
10.5	521	639	742	825	888	932	960	978	988	994
11.0	460	579	689	781	854	907	944	968	982	991
11.5	402	520	633	733	815	878	924	954	974	986
12.0	347	462	576	682	772	844	899	937	963	979
12.5	297	406	519	628	725	806	869	916	948	969
13.0	252	353	463	573	675	764	835	890	930	957
13.5	211	304	409	518	623	718	798	861	908	942
14.0	176	260	358	464	570	669	756	827	883	923
14.5	145	220	311	413	518	619	711	790	853	901
15.0	118	185	268	363	466	568	664	749	819	875

	20	21	22	23	24	25	26	27	28	29
10.5	997	999	999	1,000						
11.0	995	998	999	1,000						
11.5	992	996	998	999	1,000					
12.0	988	994	997	999	999	1,000				
12.5	983	991	995	998	999	999	1,000			
13.0	975	986	992	996	998	999	1,000			
13.5	965	980	989	994	997	998	999	1,000		
14.0	952	971	983	991	995	997	999	999	1,000	
14.5	936	960	976	986	992	996	998	999	999	1,000
15.0	917	947	967	981	989	994	997	998	999	1,000

Table 2 (continued).

c' or np' \ c	4	5	6	7	8	9	10	11	12	13
16	000	001	004	010	022	043	077	127	193	275
17	000	001	002	005	013	026	049	085	135	201
18	000	000	001	003	007	015	030	055	092	143
19	000	000	001	002	004	009	018	035	061	098
20	000	000	000	001	002	005	011	021	039	066
21	000	000	000	000	001	003	006	013	025	043
22	000	000	000	000	001	002	004	008	015	028
23	000	000	000	000	000	001	002	004	009	017
24	000	000	000	000	000	000	001	003	005	011
25	000	000	000	000	000	000	001	001	003	006

	14	15	16	17	18	19	20	21	22	23
16	368	467	566	659	742	812	868	911	942	963
17	281	371	468	564	655	736	805	861	905	937
18	208	287	375	469	562	651	731	799	855	899
19	150	215	292	378	469	561	647	725	793	849
20	105	157	221	297	381	470	559	644	721	787
21	072	111	163	227	302	384	471	558	640	716
22	048	077	117	169	232	306	387	472	556	637
23	031	052	082	123	175	238	310	389	472	555
24	020	034	056	087	128	180	243	314	392	473
25	012	022	038	060	092	134	185	247	318	394

	24	25	26	27	28	29	30	31	32	33
16	978	987	993	996	998	999	999	1,000		
17	959	975	985	991	995	997	999	999	1,000	
18	932	955	972	983	990	994	997	998	999	1,000
19	893	927	951	969	980	988	993	996	998	999
20	843	888	922	948	966	978	987	992	995	997
21	782	838	883	917	944	963	976	985	991	994
22	712	777	832	877	913	940	959	973	983	989
23	635	708	772	827	873	908	936	956	971	981
24	554	632	704	768	823	868	904	932	953	969
25	473	553	629	700	763	818	863	900	929	950

	34	35	36	37	38	39	40	41	42	43
19	999	1,000								
20	999	999	1,000							
21	997	998	999	999	1,000					
22	994	996	998	999	999	1,000				
23	988	993	996	997	999	999	1,000			
24	979	987	992	995	997	998	999	999	1,000	
25	966	978	985	991	994	997	998	999	999	1,000

Table 3   Percentiles of the *t* Distribution.

df	$t_{.60}$	$t_{.70}$	$t_{.80}$	$t_{.90}$	$t_{.95}$	$t_{.975}$	$t_{.99}$	$t_{.995}$
1	.325	.727	1.376	3.078	6.314	12.706	31.821	63.657
2	.289	.617	1.061	1.886	2.920	4.303	6.965	9.925
3	.277	.584	.978	1.638	2.353	3.182	4.541	5.841
4	.271	.569	.941	1.533	2.132	2.776	3.747	4.604
5	.267	.559	.920	1.476	2.015	2.571	3.365	4.032
6	.265	.553	.906	1.440	1.943	2.447	3.143	3.707
7	.263	.549	.896	1.415	1.895	2.365	2.998	3.499
8	.262	.546	.889	1.397	1.860	2.306	2.896	3.355
9	.261	.543	.883	1.383	1.833	2.262	2.821	3.250
10	.260	.542	.879	1.372	1.812	2.228	2.764	3.169
11	.260	.540	.876	1.363	1.796	2.201	2.718	3.106
12	.259	.539	.873	1.356	1.782	2.179	2.681	3.055
13	.259	.538	.870	1.350	1.771	2.160	2.650	3.012
14	.258	.537	.868	1.345	1.761	2.145	2.624	2.977
15	.258	.536	.866	1.341	1.753	2.131	2.602	2.947
16	.258	.535	.865	1.337	1.746	2.120	2.583	2.921
17	.257	.534	.863	1.333	1.740	2.110	2.567	2.898
18	.257	.534	.862	1.330	1.734	2.101	2.552	2.878
19	.257	.533	.861	1.328	1.729	2.093	2.539	2.861
20	.257	.533	.860	1.325	1.725	2.086	2.528	2.845
21	.257	.532	.859	1.323	1.721	2.080	2.518	2.831
22	.256	.532	.858	1.321	1.717	2.074	2.508	2.819
23	.256	.532	.858	1.319	1.714	2.069	2.500	2.807
24	.256	.531	.857	1.318	1.711	2.064	2.492	2.797
25	.256	.531	.856	1.316	1.708	2.060	2.485	2.787
26	.256	.531	.856	1.315	1.706	2.056	2.479	2.779
27	.256	.531	.855	1.314	1.703	2.052	2.473	2.771
28	.256	.530	.855	1.313	1.701	2.048	2.467	2.763
29	.256	.530	.854	1.311	1.699	2.045	2.462	2.756
30	.256	.530	.854	1.310	1.697	2.042	2.457	2.750
40	.255	.529	.851	1.303	1.684	2.021	2.423	2.704
60	.254	.527	.848	1.296	1.671	2.000	2.390	2.660
120	.254	.526	.845	1.289	1.658	1.980	2.358	2.617
$\propto$	.253	.524	.842	1.282	1.645	1.960	2.326	2.576

Adapted by permission from *Introduction to Statistical Analysis* (2d ed.) by W. J. Dixon and F. J. Massey, Jr., Copyright, 1957, McGraw-Hill Book Company, Inc. Entries originally from Table III of *Statistical Tables* by R. A. Fisher and F. Yates, 1938, Oliver and Boyd, Ltd., London.

Table 4 Percentiles of the F Distribution.

n_1 = degrees of freedom for numerator

$F_{.90}(n_1, n_2)$ F_P

n_2	1	2	3	4	5	6	7	8	9	10	12	15	20	24	30	40	60	120	∞
1	39.86	49.50	53.59	55.83	57.24	58.20	58.91	59.44	59.86	60.19	60.71	61.22	61.74	62.00	62.26	62.53	62.79	63.06	63.33
2	8.53	9.00	9.16	9.24	9.29	9.33	9.35	9.37	9.38	9.39	9.41	9.42	9.44	9.45	9.46	9.47	9.47	9.48	9.49
3	5.54	5.46	5.39	5.34	5.31	5.28	5.27	5.25	5.24	5.23	5.22	5.20	5.18	5.18	5.17	5.16	5.15	5.14	5.13
4	4.54	4.32	4.19	4.11	4.05	4.01	3.98	3.95	3.94	3.92	3.90	3.87	3.84	3.83	3.82	3.80	3.79	3.78	3.76
5	4.06	3.78	3.62	3.52	3.45	3.40	3.37	3.34	3.32	3.30	3.27	3.24	3.21	3.19	3.17	3.16	3.14	3.12	3.10
6	3.78	3.46	3.29	3.18	3.11	3.05	3.01	2.98	2.96	2.94	2.90	2.87	2.84	2.82	2.80	2.78	2.76	2.74	2.72
7	3.59	3.26	3.07	2.96	2.88	2.83	2.78	2.75	2.72	2.70	2.67	2.63	2.59	2.58	2.56	2.54	2.51	2.49	2.47
8	3.46	3.11	2.92	2.81	2.73	2.67	2.62	2.59	2.56	2.54	2.50	2.46	2.42	2.40	2.38	2.36	2.34	2.32	2.29
9	3.36	3.01	2.81	2.69	2.61	2.55	2.51	2.47	2.44	2.42	2.38	2.34	2.30	2.28	2.25	2.23	2.21	2.18	2.16
10	3.29	2.92	2.73	2.61	2.52	2.46	2.41	2.38	2.35	2.32	2.28	2.24	2.20	2.18	2.16	2.13	2.11	2.08	2.06
11	3.23	2.86	2.66	2.54	2.45	2.39	2.34	2.30	2.27	2.25	2.21	2.17	2.12	2.10	2.08	2.05	2.03	2.00	1.97
12	3.18	2.81	2.61	2.48	2.39	2.33	2.28	2.24	2.21	2.19	2.15	2.10	2.06	2.04	2.01	1.99	1.96	1.93	1.90
13	3.14	2.76	2.56	2.43	2.35	2.28	2.23	2.20	2.16	2.14	2.10	2.05	2.01	1.98	1.96	1.93	1.90	1.88	1.85
14	3.10	2.73	2.52	2.39	2.31	2.24	2.19	2.15	2.12	2.10	2.05	2.01	1.96	1.94	1.91	1.89	1.86	1.83	1.80
15	3.07	2.70	2.49	2.36	2.27	2.21	2.16	2.12	2.09	2.06	2.02	1.97	1.92	1.90	1.87	1.85	1.82	1.79	1.76
16	3.05	2.67	2.46	2.33	2.24	2.18	2.13	2.09	2.06	2.03	1.99	1.94	1.89	1.87	1.84	1.81	1.78	1.75	1.72
17	3.03	2.64	2.44	2.31	2.22	2.15	2.10	2.06	2.03	2.00	1.96	1.91	1.86	1.84	1.81	1.78	1.75	1.72	1.69
18	3.01	2.62	2.42	2.29	2.20	2.13	2.08	2.04	2.00	1.98	1.93	1.89	1.84	1.81	1.78	1.75	1.72	1.69	1.66
19	2.99	2.61	2.40	2.27	2.18	2.11	2.06	2.02	1.98	1.96	1.91	1.86	1.81	1.79	1.76	1.73	1.70	1.67	1.63
20	2.97	2.59	2.38	2.25	2.16	2.09	2.04	2.00	1.96	1.94	1.89	1.84	1.79	1.77	1.74	1.71	1.68	1.64	1.61
21	2.96	2.57	2.36	2.23	2.14	2.08	2.02	1.98	1.95	1.92	1.87	1.83	1.78	1.75	1.72	1.69	1.66	1.62	1.59
22	2.95	2.56	2.35	2.22	2.13	2.06	2.01	1.97	1.93	1.90	1.86	1.81	1.76	1.73	1.70	1.67	1.64	1.60	1.57
23	2.94	2.55	2.34	2.21	2.11	2.05	1.99	1.95	1.92	1.89	1.84	1.80	1.74	1.72	1.69	1.66	1.62	1.59	1.55
24	2.93	2.54	2.33	2.19	2.10	2.04	1.98	1.94	1.91	1.88	1.83	1.78	1.73	1.70	1.67	1.64	1.61	1.57	1.53
25	2.92	2.53	2.32	2.18	2.09	2.02	1.97	1.93	1.89	1.87	1.82	1.77	1.72	1.69	1.66	1.63	1.59	1.56	1.52
26	2.91	2.52	2.31	2.17	2.08	2.01	1.96	1.92	1.88	1.86	1.81	1.76	1.71	1.68	1.65	1.61	1.58	1.54	1.50
27	2.90	2.51	2.30	2.17	2.07	2.00	1.95	1.91	1.87	1.85	1.80	1.75	1.70	1.67	1.64	1.60	1.57	1.53	1.49
28	2.89	2.50	2.29	2.16	2.06	2.00	1.94	1.90	1.87	1.84	1.79	1.74	1.69	1.66	1.63	1.59	1.56	1.52	1.48
29	2.89	2.50	2.28	2.15	2.06	1.99	1.93	1.89	1.86	1.83	1.78	1.73	1.68	1.65	1.62	1.58	1.55	1.51	1.47
30	2.88	2.49	2.28	2.14	2.05	1.98	1.93	1.88	1.85	1.82	1.77	1.72	1.67	1.64	1.61	1.57	1.54	1.50	1.46
40	2.84	2.44	2.23	2.09	2.00	1.93	1.87	1.83	1.79	1.76	1.71	1.66	1.61	1.57	1.54	1.51	1.47	1.42	1.38
60	2.79	2.39	2.18	2.04	1.95	1.87	1.82	1.77	1.74	1.71	1.66	1.60	1.54	1.51	1.48	1.44	1.40	1.35	1.29
120	2.75	2.35	2.13	1.99	1.90	1.82	1.77	1.72	1.68	1.65	1.60	1.55	1.48	1.45	1.41	1.37	1.32	1.26	1.19
∞	2.71	2.30	2.08	1.94	1.85	1.77	1.72	1.67	1.63	1.60	1.55	1.49	1.42	1.38	1.34	1.30	1.24	1.17	1.00

n_2 = degrees of freedom for denominator

218

$F_{.95}$ (n_1, n_2)

n_1 = degrees of freedom for numerator

n_2	1	2	3	4	5	6	7	8	9	10	12	15	20	24	30	40	60	120	∞
1	161.4	199.5	215.7	224.6	230.2	234.0	236.8	238.9	240.5	241.9	243.9	245.9	248.0	249.1	250.1	251.1	252.2	253.3	254.3
2	18.51	19.00	19.16	19.25	19.30	19.33	19.35	19.37	19.38	19.40	19.41	19.43	19.45	19.45	19.46	19.47	19.48	19.49	19.50
3	10.13	9.55	9.28	9.12	9.01	8.94	8.89	8.85	8.81	8.79	8.74	8.70	8.66	8.64	8.62	8.59	8.57	8.55	8.53
4	7.71	6.94	6.59	6.39	6.26	6.16	6.09	6.04	6.00	5.96	5.91	5.86	5.80	5.77	5.75	5.72	5.69	5.66	5.63
5	6.61	5.79	5.41	5.19	5.05	4.95	4.88	4.82	4.77	4.74	4.68	4.62	4.56	4.53	4.50	4.46	4.43	4.40	4.36
6	5.99	5.14	4.76	4.53	4.39	4.28	4.21	4.15	4.10	4.06	4.00	3.94	3.87	3.84	3.81	3.77	3.74	3.70	3.67
7	5.59	4.74	4.35	4.12	3.97	3.87	3.79	3.73	3.68	3.64	3.57	3.51	3.44	3.41	3.38	3.34	3.30	3.27	3.23
8	5.32	4.46	4.07	3.84	3.69	3.58	3.50	3.44	3.39	3.35	3.28	3.22	3.15	3.12	3.08	3.04	3.01	2.97	2.93
9	5.12	4.26	3.86	3.63	3.48	3.37	3.29	3.23	3.18	3.14	3.07	3.01	2.94	2.90	2.86	2.83	2.79	2.75	2.71
10	4.96	4.10	3.71	3.48	3.33	3.22	3.14	3.07	3.02	2.98	2.91	2.85	2.77	2.74	2.70	2.66	2.62	2.58	2.54
11	4.84	3.98	3.59	3.36	3.20	3.09	3.01	2.95	2.90	2.85	2.79	2.72	2.65	2.61	2.57	2.53	2.49	2.45	2.40
12	4.75	3.89	3.49	3.26	3.11	3.00	2.91	2.85	2.80	2.75	2.69	2.62	2.54	2.51	2.47	2.43	2.38	2.34	2.30
13	4.67	3.81	3.41	3.18	3.03	2.92	2.83	2.77	2.71	2.67	2.60	2.53	2.46	2.42	2.38	2.34	2.30	2.25	2.21
14	4.60	3.74	3.34	3.11	2.96	2.85	2.76	2.70	2.65	2.60	2.53	2.46	2.39	2.35	2.31	2.27	2.22	2.18	2.13
15	4.54	3.68	3.29	3.06	2.90	2.79	2.71	2.64	2.59	2.54	2.48	2.40	2.33	2.29	2.25	2.20	2.16	2.11	2.07
16	4.49	3.63	3.24	3.01	2.85	2.74	2.66	2.59	2.54	2.49	2.42	2.35	2.28	2.24	2.19	2.15	2.11	2.06	2.01
17	4.45	3.59	3.20	2.96	2.81	2.70	2.61	2.55	2.49	2.45	2.38	2.31	2.23	2.19	2.15	2.10	2.06	2.01	1.96
18	4.41	3.55	3.16	2.93	2.77	2.66	2.58	2.51	2.46	2.41	2.34	2.27	2.19	2.15	2.11	2.06	2.02	1.97	1.92
19	4.38	3.52	3.13	2.90	2.74	2.63	2.54	2.48	2.42	2.38	2.31	2.23	2.16	2.11	2.07	2.03	1.98	1.93	1.88
20	4.35	3.49	3.10	2.87	2.71	2.60	2.51	2.45	2.39	2.35	2.28	2.20	2.12	2.08	2.04	1.99	1.95	1.90	1.84
21	4.32	3.47	3.07	2.84	2.68	2.57	2.49	2.42	2.37	2.32	2.25	2.18	2.10	2.05	2.01	1.96	1.92	1.87	1.81
22	4.30	3.44	3.05	2.82	2.66	2.55	2.46	2.40	2.34	2.30	2.23	2.15	2.07	2.03	1.98	1.94	1.89	1.84	1.78
23	4.28	3.42	3.03	2.80	2.64	2.53	2.44	2.37	2.32	2.27	2.20	2.13	2.05	2.01	1.96	1.91	1.86	1.81	1.76
24	4.26	3.40	3.01	2.78	2.62	2.51	2.42	2.36	2.30	2.25	2.18	2.11	2.03	1.98	1.94	1.89	1.84	1.79	1.73
25	4.24	3.39	2.99	2.76	2.60	2.49	2.40	2.34	2.28	2.24	2.16	2.09	2.01	1.96	1.92	1.87	1.82	1.77	1.71
26	4.23	3.37	2.98	2.74	2.59	2.47	2.39	2.32	2.27	2.22	2.15	2.07	1.99	1.95	1.90	1.85	1.80	1.75	1.69
27	4.21	3.35	2.96	2.73	2.57	2.46	2.37	2.31	2.25	2.20	2.13	2.06	1.97	1.93	1.88	1.84	1.79	1.73	1.67
28	4.20	3.34	2.95	2.71	2.56	2.45	2.36	2.29	2.24	2.19	2.12	2.04	1.96	1.91	1.87	1.82	1.77	1.71	1.65
29	4.18	3.33	2.93	2.70	2.55	2.43	2.35	2.28	2.22	2.18	2.10	2.03	1.94	1.90	1.85	1.81	1.75	1.70	1.64
30	4.17	3.32	2.92	2.69	2.53	2.42	2.33	2.27	2.21	2.16	2.09	2.01	1.93	1.89	1.84	1.79	1.74	1.68	1.62
40	4.08	3.23	2.84	2.61	2.45	2.34	2.25	2.18	2.12	2.08	2.00	1.92	1.84	1.79	1.74	1.69	1.64	1.58	1.51
60	4.00	3.15	2.76	2.53	2.37	2.25	2.17	2.10	2.04	1.99	1.92	1.84	1.75	1.70	1.65	1.59	1.53	1.47	1.39
120	3.92	3.07	2.68	2.45	2.29	2.17	2.09	2.02	1.96	1.91	1.83	1.75	1.66	1.61	1.55	1.50	1.43	1.35	1.25
∞	3.84	3.00	2.60	2.37	2.21	2.10	2.01	1.94	1.88	1.83	1.75	1.67	1.57	1.52	1.46	1.39	1.32	1.22	1.00

n_2 = degrees of freedom for denominator

Adapted with permission from *Biometrika Tables for Statisticians*, Vol. I (2d ed.), edited by E. S. Pearson and H. O. Hartley, Copyright 1958, Cambridge University Press.

Table 4 (continued).

$F_{.975}(n_1, n_2)$

n_1 = degrees of freedom for numerator

$n_2 \backslash n_1$	1	2	3	4	5	6	7	8	9	10	12	15	20	24	30	40	60	120	∞
1	647.8	799.5	864.2	899.6	921.8	937.1	948.2	956.7	963.3	968.6	976.7	984.9	993.1	997.2	1001	1006	1010	1014	1018
2	38.51	39.00	39.17	39.25	39.30	39.33	39.36	39.37	39.39	39.40	39.41	39.43	39.45	39.46	39.46	39.47	39.48	39.49	39.50
3	17.44	16.04	15.44	15.10	14.88	14.73	14.62	14.54	14.47	14.42	14.34	14.25	14.17	14.12	14.08	14.04	13.99	13.95	13.90
4	12.22	10.65	9.98	9.60	9.36	9.20	9.07	8.98	8.90	8.84	8.75	8.66	8.56	8.51	8.46	8.41	8.36	8.31	8.26
5	10.01	8.43	7.76	7.39	7.15	6.98	6.85	6.76	6.68	6.62	6.52	6.43	6.33	6.28	6.23	6.18	6.12	6.07	6.02
6	8.81	7.26	6.60	6.23	5.99	5.82	5.70	5.60	5.52	5.46	5.37	5.27	5.17	5.12	5.07	5.01	4.96	4.90	4.85
7	8.07	6.54	5.89	5.52	5.29	5.12	4.99	4.90	4.82	4.76	4.67	4.57	4.47	4.42	4.36	4.31	4.25	4.20	4.14
8	7.57	6.06	5.42	5.05	4.82	4.65	4.53	4.43	4.36	4.30	4.20	4.10	4.00	3.95	3.89	3.84	3.78	3.73	3.67
9	7.21	5.71	5.08	4.72	4.48	4.32	4.20	4.10	4.03	3.96	3.87	3.77	3.67	3.61	3.56	3.51	3.45	3.39	3.33
10	6.94	5.46	4.83	4.47	4.24	4.07	3.95	3.85	3.78	3.72	3.62	3.52	3.42	3.37	3.31	3.26	3.20	3.14	3.08
11	6.72	5.26	4.63	4.28	4.04	3.88	3.76	3.66	3.59	3.53	3.43	3.33	3.23	3.17	3.12	3.06	3.00	2.94	2.88
12	6.55	5.10	4.47	4.12	3.89	3.73	3.61	3.51	3.44	3.37	3.28	3.18	3.07	3.02	2.96	2.91	2.85	2.79	2.72
13	6.41	4.97	4.35	4.00	3.77	3.60	3.48	3.39	3.31	3.25	3.15	3.05	2.95	2.89	2.84	2.78	2.72	2.66	2.60
14	6.30	4.86	4.24	3.89	3.66	3.50	3.38	3.29	3.21	3.15	3.05	2.95	2.84	2.79	2.73	2.67	2.61	2.55	2.49
15	6.20	4.77	4.15	3.80	3.58	3.41	3.29	3.20	3.12	3.06	2.96	2.86	2.76	2.70	2.64	2.59	2.52	2.46	2.40
16	6.12	4.69	4.08	3.73	3.50	3.34	3.22	3.12	3.05	2.99	2.89	2.79	2.68	2.63	2.57	2.51	2.45	2.38	2.32
17	6.04	4.62	4.01	3.66	3.44	3.28	3.16	3.06	2.98	2.92	2.82	2.72	2.62	2.56	2.50	2.44	2.38	2.32	2.25
18	5.98	4.56	3.95	3.61	3.38	3.22	3.10	3.01	2.93	2.87	2.77	2.67	2.56	2.50	2.44	2.38	2.32	2.26	2.19
19	5.92	4.51	3.90	3.56	3.33	3.17	3.05	2.96	2.88	2.82	2.72	2.62	2.51	2.45	2.39	2.33	2.27	2.20	2.13
20	5.87	4.46	3.86	3.51	3.29	3.13	3.01	2.91	2.84	2.77	2.68	2.57	2.46	2.41	2.35	2.29	2.22	2.16	2.09
21	5.83	4.42	3.82	3.48	3.25	3.09	2.97	2.87	2.80	2.73	2.64	2.53	2.42	2.37	2.31	2.25	2.18	2.11	2.04
22	5.79	4.38	3.78	3.44	3.22	3.05	2.93	2.84	2.76	2.70	2.60	2.50	2.39	2.33	2.27	2.21	2.14	2.08	2.00
23	5.75	4.35	3.75	3.41	3.18	3.02	2.90	2.81	2.73	2.67	2.57	2.47	2.36	2.30	2.24	2.18	2.11	2.04	1.97
24	5.72	4.32	3.72	3.38	3.15	2.99	2.87	2.78	2.70	2.64	2.54	2.44	2.33	2.27	2.21	2.15	2.08	2.01	1.94
25	5.69	4.29	3.69	3.35	3.13	2.97	2.85	2.75	2.68	2.61	2.51	2.41	2.30	2.24	2.18	2.12	2.05	1.98	1.91
26	5.66	4.27	3.67	3.33	3.10	2.94	2.82	2.73	2.65	2.59	2.49	2.39	2.28	2.22	2.16	2.09	2.03	1.95	1.88
27	5.63	4.24	3.65	3.31	3.08	2.92	2.80	2.71	2.63	2.57	2.47	2.36	2.25	2.19	2.13	2.07	2.00	1.93	1.85
28	5.61	4.22	3.63	3.29	3.06	2.90	2.78	2.69	2.61	2.55	2.45	2.34	2.23	2.17	2.11	2.05	1.98	1.91	1.83
29	5.59	4.20	3.61	3.27	3.04	2.88	2.76	2.67	2.59	2.53	2.43	2.32	2.21	2.15	2.09	2.03	1.96	1.89	1.81
30	5.57	4.18	3.59	3.25	3.03	2.87	2.75	2.65	2.57	2.51	2.41	2.31	2.20	2.14	2.07	2.01	1.94	1.87	1.79
40	5.42	4.05	3.46	3.13	2.90	2.74	2.62	2.53	2.45	2.39	2.29	2.18	2.07	2.01	1.94	1.88	1.80	1.72	1.64
60	5.29	3.93	3.34	3.01	2.79	2.63	2.51	2.41	2.33	2.27	2.17	2.06	1.94	1.88	1.82	1.74	1.67	1.58	1.48
120	5.15	3.80	3.23	2.89	2.67	2.52	2.39	2.30	2.22	2.16	2.05	1.94	1.82	1.76	1.69	1.61	1.53	1.43	1.31
∞	5.02	3.69	3.12	2.79	2.57	2.41	2.29	2.19	2.11	2.05	1.94	1.83	1.71	1.64	1.57	1.48	1.39	1.27	1.00

n_2 = degrees of freedom for denominator

F.99 (n₁, n₂)

n_1 = degrees of freedom for numerator

n_2	1	2	3	4	5	6	7	8	9	10	12	15	20	24	30	40	60	120	∞
1	4052	4999.5	5403	5625	5764	5859	5928	5982	6022	6056	6106	6157	6209	6235	6261	6287	6313	6339	6366
2	98.50	99.00	99.17	99.25	99.30	99.33	99.36	99.37	99.39	99.40	99.42	99.43	99.45	99.46	99.47	99.47	99.48	99.49	99.50
3	34.12	30.82	29.46	28.71	28.24	27.91	27.67	27.49	27.35	27.23	27.05	26.87	26.69	26.60	26.50	26.41	26.32	26.22	26.13
4	21.20	18.00	16.69	15.98	15.52	15.21	14.98	14.80	14.66	14.55	14.37	14.20	14.02	13.93	13.84	13.75	13.65	13.56	13.46
5	16.26	13.27	12.06	11.39	10.97	10.67	10.46	10.29	10.16	10.05	9.89	9.72	9.55	9.47	9.38	9.29	9.20	9.11	9.02
6	13.75	10.92	9.78	9.15	8.75	8.47	8.26	8.10	7.98	7.87	7.72	7.56	7.40	7.31	7.23	7.14	7.06	6.97	6.88
7	12.25	9.55	8.45	7.85	7.46	7.19	6.99	6.84	6.72	6.62	6.47	6.31	6.16	6.07	5.99	5.91	5.82	5.74	5.65
8	11.26	8.65	7.59	7.01	6.63	6.37	6.18	6.03	5.91	5.81	5.67	5.52	5.36	5.28	5.20	5.12	5.03	4.95	4.86
9	10.56	8.02	6.99	6.42	6.06	5.80	5.61	5.47	5.35	5.26	5.11	4.96	4.81	4.73	4.65	4.57	4.48	4.40	4.31
10	10.04	7.56	6.55	5.99	5.64	5.39	5.20	5.06	4.94	4.85	4.71	4.56	4.41	4.33	4.25	4.17	4.08	4.00	3.91
11	9.65	7.21	6.22	5.67	5.32	5.07	4.89	4.74	4.63	4.54	4.40	4.25	4.10	4.02	3.94	3.86	3.78	3.69	3.60
12	9.33	6.93	5.95	5.41	5.06	4.82	4.64	4.50	4.39	4.30	4.16	4.01	3.86	3.78	3.70	3.62	3.54	3.45	3.36
13	9.07	6.70	5.74	5.21	4.86	4.62	4.44	4.30	4.19	4.10	3.96	3.82	3.66	3.59	3.51	3.43	3.34	3.25	3.17
14	8.86	6.51	5.56	5.04	4.69	4.46	4.28	4.14	4.03	3.94	3.80	3.66	3.51	3.43	3.35	3.27	3.18	3.09	3.00
15	8.68	6.36	5.42	4.89	4.56	4.32	4.14	4.00	3.89	3.80	3.67	3.52	3.37	3.29	3.21	3.13	3.05	2.96	2.87
16	8.53	6.23	5.29	4.77	4.44	4.20	4.03	3.89	3.78	3.69	3.55	3.41	3.26	3.18	3.10	3.02	2.93	2.84	2.75
17	8.40	6.11	5.18	4.67	4.34	4.10	3.93	3.79	3.68	3.59	3.46	3.31	3.16	3.08	3.00	2.92	2.83	2.75	2.65
18	8.29	6.01	5.09	4.58	4.25	4.01	3.84	3.71	3.60	3.51	3.37	3.23	3.08	3.00	2.92	2.84	2.75	2.66	2.57
19	8.18	5.93	5.01	4.50	4.17	3.94	3.77	3.63	3.52	3.43	3.30	3.15	3.00	2.92	2.84	2.76	2.67	2.58	2.49
20	8.10	5.85	4.94	4.43	4.10	3.87	3.70	3.56	3.46	3.37	3.23	3.09	2.94	2.86	2.78	2.69	2.61	2.52	2.42
21	8.02	5.78	4.87	4.37	4.04	3.81	3.64	3.51	3.40	3.31	3.17	3.03	2.88	2.80	2.72	2.64	2.55	2.46	2.36
22	7.95	5.72	4.82	4.31	3.99	3.76	3.59	3.45	3.35	3.26	3.12	2.98	2.83	2.75	2.67	2.58	2.50	2.40	2.31
23	7.88	5.66	4.76	4.26	3.94	3.71	3.54	3.41	3.30	3.21	3.07	2.93	2.78	2.70	2.62	2.54	2.45	2.35	2.26
24	7.82	5.61	4.72	4.22	3.90	3.67	3.50	3.36	3.26	3.17	3.03	2.89	2.74	2.66	2.58	2.49	2.40	2.31	2.21
25	7.77	5.57	4.68	4.18	3.85	3.63	3.46	3.32	3.22	3.13	2.99	2.85	2.70	2.62	2.54	2.45	2.36	2.27	2.17
26	7.72	5.53	4.64	4.14	3.82	3.59	3.42	3.29	3.18	3.09	2.96	2.81	2.66	2.58	2.50	2.42	2.33	2.23	2.13
27	7.68	5.49	4.60	4.11	3.78	3.56	3.39	3.26	3.15	3.06	2.93	2.78	2.63	2.55	2.47	2.38	2.29	2.20	2.10
28	7.64	5.45	4.57	4.07	3.75	3.53	3.36	3.23	3.12	3.03	2.90	2.75	2.60	2.52	2.44	2.35	2.26	2.17	2.06
29	7.60	5.42	4.54	4.04	3.73	3.50	3.33	3.20	3.09	3.00	2.87	2.73	2.57	2.49	2.41	2.33	2.23	2.14	2.03
30	7.56	5.39	4.51	4.02	3.70	3.47	3.30	3.17	3.07	2.98	2.84	2.70	2.55	2.47	2.39	2.30	2.21	2.11	2.01
40	7.31	5.18	4.31	3.83	3.51	3.29	3.12	2.99	2.89	2.80	2.66	2.52	2.37	2.29	2.20	2.11	2.02	1.92	1.80
60	7.08	4.98	4.13	3.65	3.34	3.12	2.95	2.82	2.72	2.63	2.50	2.35	2.20	2.12	2.03	1.94	1.84	1.73	1.60
120	6.85	4.79	3.95	3.48	3.17	2.96	2.79	2.66	2.56	2.47	2.34	2.19	2.03	1.95	1.86	1.76	1.66	1.53	1.38
∞	6.63	4.61	3.78	3.32	3.02	2.80	2.64	2.51	2.41	2.32	2.18	2.04	1.88	1.79	1.70	1.59	1.47	1.32	1.00

n_2 = degrees of freedom for denominator

Table 5   Percentiles of the x^2 Distribution.

Values of χ^2_P corresponding to P

df	$\chi^2_{.005}$	$\chi^2_{.01}$	$\chi^2_{.025}$	$\chi^2_{.05}$	$\chi^2_{.10}$	$\chi^2_{.90}$	$\chi^2_{.95}$	$\chi^2_{.975}$	$\chi^2_{.99}$	$\chi^2_{.995}$
1	.000039	.00016	.00098	.0039	.0158	2.71	3.84	5.02	6.63	7.88
2	.0100	.0201	.0506	.1026	.2107	4.61	5.99	7.38	9.21	10.60
3	.0717	.115	.216	.352	.584	6.25	7.81	9.35	11.34	12.84
4	.207	.297	.484	.711	1.064	7.78	9.49	11.14	13.28	14.86
5	.412	.554	.831	1.15	1.61	9.24	11.07	12.83	15.09	16.75
6	.676	.872	1.24	1.64	2.20	10.64	12.59	14.45	16.81	18.55
7	.989	1.24	1.69	2.17	2.83	12.02	14.07	16.01	18.48	20.28
8	1.34	1.65	2.18	2.73	3.49	13.36	15.51	17.53	20.09	21.96
9	1.73	2.09	2.70	3.33	4.17	14.68	16.92	19.02	21.67	23.59
10	2.16	2.56	3.25	3.94	4.87	15.99	18.31	20.48	23.21	25.19
11	2.60	3.05	3.82	4.57	5.58	17.28	19.68	21.92	24.73	26.76
12	3.07	3.57	4.40	5.23	6.30	18.55	21.03	23.34	26.22	28.30
13	3.57	4.11	5.01	5.89	7.04	19.81	22.36	24.74	27.69	29.82
14	4.07	4.66	5.63	6.57	7.79	21.06	23.68	26.12	29.14	31.32
15	4.60	5.23	6.26	7.26	8.55	22.31	25.00	27.49	30.58	32.80
16	5.14	5.81	6.91	7.96	9.31	23.54	26.30	28.85	32.00	34.27
18	6.26	7.01	8.23	9.39	10.86	25.99	28.87	31.53	34.81	37.16
20	7.43	8.26	9.59	10.85	12.44	28.41	31.41	34.17	37.57	40.00
24	9.89	10.86	12.40	13.85	15.66	33.20	36.42	39.36	42.98	45.56
30	13.79	14.95	16.79	18.49	20.60	40.26	43.77	46.98	50.89	53.67
40	20.71	22.16	24.43	26.51	29.05	51.81	55.76	59.34	63.69	66.77
60	35.53	37.48	40.48	43.19	46.46	74.40	79.08	83.30	88.38	91.95
120	83.85	86.92	91.58	95.70	100.62	140.23	146.57	152.21	158.95	163.64

For large degrees of freedom,

$$\chi^2_P = \tfrac{1}{2} (z_P + \sqrt{2\nu - 1})^2 \text{ approximately,}$$

where ν = degrees of freedom and z_P is given in Table A-2.

Adapted with permission from *Introduction to Statistical Analysis* (2d ed.) by W. J. Dixon and F. J. Massey, Jr., Copyright, 1957, McGraw-Hill Book Company, Inc.

Table 6 Factors for Control Charts for Variables X, Y, s, R: Normal Universe Factors for Computing Central Lines and 3-Sigma Control Limits

Observations in Sample, n	Chart for Averages			Chart for Standard Deviations						Chart for Ranges						
	Factors for Control Limits			Factors for Central Line		Factors for Control Limits				Factors for Central Line			Factors for Control Limits			
	A	A_2	A_3	c_4	$1/c_4$	B_3	B_4	B_5	B_6	d_2	$1/d_2$	d_3	D_1	D_2	D_3	D_4
2	2.121	1.880	2.659	0.7979	1.2533	0	3.267	0	2.606	1.128	0.8865	0.853	0	3.686	0	3.267
3	1.732	1.023	1.954	0.8862	1.1284	0	2.568	0	2.276	1.693	0.5907	0.888	0	4.358	0	2.574
4	1.500	0.729	1.628	0.9213	1.0854	0	2.266	0	2.088	2.059	0.4857	0.880	0	4.698	0	2.282
5	1.342	0.577	1.427	0.9400	1.0638	0	2.089	0	1.964	2.326	0.4299	0.864	0	4.918	0	2.114
6	1.225	0.483	1.287	0.9515	1.0510	0.030	1.970	0.029	1.874	2.534	0.3946	0.848	0	5.078	0	2.004
7	1.134	0.419	1.182	0.9594	1.0423	0.118	1.882	0.113	1.806	2.704	0.3698	0.833	0.204	5.204	0.076	1.924
8	1.061	0.373	1.099	0.9650	1.0363	0.185	1.815	0.179	1.751	2.847	0.3512	0.820	0.388	5.306	0.136	1.864
9	1.000	0.337	1.032	0.9693	1.0317	0.239	1.761	0.232	1.707	2.970	0.3367	0.808	0.547	5.393	0.184	1.816
10	0.949	0.308	0.975	0.9727	1.0281	0.284	1.716	0.276	1.669	3.078	0.3249	0.797	0.687	5.469	0.223	1.777
11	0.905	0.285	0.927	0.9754	1.0252	0.321	1.679	0.313	1.637	3.173	0.3152	0.787	0.811	5.535	0.256	1.744
12	0.866	0.266	0.886	0.9776	1.0229	0.354	1.646	0.346	1.610	3.258	0.3069	0.778	0.922	5.594	0.283	1.717
13	0.832	0.249	0.850	0.9794	1.0210	0.382	1.618	0.374	1.585	3.336	0.2998	0.770	1.025	5.647	0.307	1.693
14	0.802	0.235	0.817	0.9810	1.0194	0.406	1.594	0.399	1.563	3.407	0.2935	0.763	1.118	5.696	0.328	1.672
15	0.775	0.223	0.789	0.9823	1.0180	0.428	1.572	0.421	1.544	3.472	0.2880	0.756	1.203	5.741	0.347	1.653
16	0.750	0.212	0.763	0.9835	1.0168	0.448	1.552	0.440	1.526	3.532	0.2831	0.750	1.282	5.782	0.363	1.637
17	0.728	0.203	0.739	0.9845	1.0157	0.466	1.534	0.458	1.511	3.588	0.2787	0.744	1.356	5.820	0.378	1.622
18	0.707	0.194	0.718	0.9854	1.0148	0.482	1.518	0.475	1.496	3.640	0.2747	0.739	1.424	5.856	0.391	1.608
19	0.688	0.187	0.698	0.9862	1.0140	0.497	1.503	0.490	1.483	3.689	0.2711	0.734	1.487	5.891	0.403	1.597
20	0.671	0.180	0.680	0.9869	1.0133	0.510	1.490	0.504	1.470	3.735	0.2677	0.729	1.549	5.921	0.415	1.585
21	0.655	0.173	0.663	0.9876	1.0126	0.523	1.477	0.516	1.459	3.778	0.2647	0.724	1.605	5.951	0.425	1.575
22	0.640	0.167	0.647	0.9882	1.0119	0.534	1.466	0.528	1.448	3.819	0.2618	0.720	1.659	5.979	0.434	1.566
23	0.626	0.162	0.633	0.9887	1.0114	0.545	1.455	0.539	1.438	3.858	0.2592	0.716	1.710	6.006	0.443	1.557
24	0.612	0.157	0.619	0.9892	1.0109	0.555	1.445	0.549	1.429	3.895	0.2567	0.712	1.759	6.031	0.451	1.548
25	0.600	0.153	0.606	0.9896	1.0105	0.565	1.435	0.559	1.420	3.931	0.2544	0.708	1.806	6.056	0.459	1.541

Table reproduced from ASTM-STP 15D by kind permission of the American Society for Testing and Materials.

Table 7   Table for values of "y," to be used in calculating
AOQL.

c	0	1	2
y	0.368	0.841	1.372
c	3	4	5
y	1.946	2.544	3.172
c	6	7	8
y	3.810	4.465	5.150
c	9	10	11
y	5.836	6.535	7.234

References

American Society for Quality Control. General Requirements for a Quality Program. ASQC Standard C1-1968. Milwaukee, Wis. 1968.

American Society for Quality Control. Generic Guidelines for Quality Systems. ANSI/ASQC Z1.15-1979. Milwaukee, Wis. 1979.

Bennett, A., and Solomon, J. Quality Award Becoming Job 1 for U.S. Companies? *Wall Street Journal,* Nov. 4, 1989.

Brainard, E.H. Just How Good Are Vendor Surveys? *Quality Assurance,* pp. 22-25, August 1966.

Carlson, R.D., Gerber, J., and McHugh, J.F. *Manual of Quality Assurance Procedures and Forms.* Englewood Cliffs, N.J.: Prentice-Hall, 1981.

Castro, J. Making It Better, *Time,* Nov. 13, 1989.

Crosby, P.B. *Quality Without Tears.* New York: Plume Books, 1984.

Crosby, P.B. *Quality is Free.* New York: McGraw-Hill, 1979.

DeCario, N.J., and Sterett, W.K. History of the Malcolm Baldrige National Quality Award, *Quality Progress,* March 1990.

Deming, W.E. *Quality, Productivity, and Competitive Position,* Cambridge, Mass.: Center for Advanced Engineering Study, M.I.T., 1982.

Deming, W.E. *Out of the Crisis,* Cambridge, Mass.: Center for Advanced Engineering Study, M.I.T., 1986.

DoD, *Total Quality Management Master Plan,* Washington, D.C.: Department of Defense, August, 1988.

DoD, *An Education and Training Strategy for Total Quality Management in the Department of Defense,* Washington, D.C.: Department of Defense, July, 1989.

DoD, *Total Quality Management Guide, Volumes 1 and 2, DoD 5000.51-G,* Washington, D.C.: Department of Defense, Feb., 1990.

Duncan, A.J. *Quality Control and Industrial Statistics, 4th ed.* Homewood, Ill.: Richard D. Irwin, 1974.

Feigenbaum, A.V. *Total Quality Control,* 3rd ed. New York: McGraw-Hill, 1983.

Field, D.L., ed. *Procurement Quality Control: A Handbook of Recommended Practices,* 2nd ed. Milwaukee, Wis.: American Society for Quality Control, 1976.

Fortuna, R.M., Beyond Quality: Taking SPC Upstream, *Quality Progress,* June, 1988.

Fukuda, R. *Managerial Engineering.* Stamford, Conn.: Productivity Inc., 1983.

Gibra, I.N. Economically Optimal Determination of Parameters of np Control Charts, *Journal of Quality Technology,* vol. 10, no. 1, pp. 12-19, 1978.

Gibb, C. "TQM and the Quality Professional," *The Quality Management Forum, ASQC Management Division Newsletter,* Winter, 1990.

Gitlow, H.S., and Gitlow, S.J. *The Deming Guide to Quality and Competitive Position.* Englwood Cliffs, N.J.: Prentice-Hall, 1987.

Hayes, G.E., and Romig, H.G. *Modern Quality Control,* rev. ed. Encino, Calif.: Glencoe, 1982.

Hayes, G. Three Views of TQM, *Quality Magazine,* April, 1990.

Hillkirk, J. Bush Hands Out Baldrige Awards, *USA Today,* Nov. 3, 1989.

Ishikawa, K. *What is Total Quality Control the Japanese Way?* Englewood Cliffs, N.J.: Prentice-Hall, 1985.

Juran, J.M. and Gryna, F.M. Jr., *Quality Planning and Analysis,* 2nd ed. New York: McGraw-Hill, 1980.

Juran, J.M., ed. *Quality Control Handbook,* 3rd ed. New York: McGraw-Hill, 1979.

Kackar, R.N. Off-Line Quality Control, Parameter Design, and the Taguchi Method, *Journal of Quality Technology,* vol. 17, no. 4, pp. 176-188, 1985.

Kivenko, K. *Quality Control for Management.* Englewood Cliffs, N.J.: Prentice-Hall, 1984.

King, B. *Better Designs in Half the Time: Implementing QFD in America,* Methuen, Mass.: Goal/QPC, 1987.

McGreggor, D. *The Human Side of Enterprise.* New York: McGraw-Hill, 1960.

Mizuno, S. *Management for Quality Improvement: The Seven New QC Tools.* Cambridge, Mass.: Productivity Press, 1988.

Montgomery, D.C. *Design and Analysis of Experiments,* 2nd ed. New York: Wiley, 1984.

Montgomery, D.C. *Introduction to Statistical Quality Control.* New York: Wiley, 1986.

Nelson, L.S. The Shewhart Control Chart—Tests for Special Causes, *Journal of Quality Technology,* vol. 16, no. 4, pp. 237-239, 1984.

Pyzdek, T. *An SPC Primer: Programmed Learning Guide to Statistical Process Control.* Tucson, Ariz.: Quality America, 1984.

Pyzdek, T. *Certified Quality Engineer Examination Study Guide.* Tucson, Ariz.: Quality America, 1986.

Pyzdek, T. The Impact of Quality Cost Reductions on Profits, *Quality Progress,* pp. 14-15, November 1976.

Pyzdek, T. *Pyzdek's Guide to SPC.* Tucson, Ariz.: Quality America, 1989.

Quest for Quality, *Electronics Business Magazine Special Issue,* vol. 15, no. 20, October 16, 1989, Cahners Publishing Company, Newton, MA.

ReVelle, J.B. *The New Quality Technology.* Los Angeles, Calif.: Hughes Aircraft Co. (213), pp. 568-6202, 1988.

Scherkenbach, W.W. *The Deming Route to Quality and Productivity.* Washington, D.C.: Ceep Press, 1986.

Schonberger, R.J. *Japanese Manufacturing Techniques.* New York: Free Press, 1982.

Shewhart, W.A. *Economic Control oof Quality of Manufactured Product.* New York: Van Nostrand, 1931. Reprinted by the American Society for Quality Control, Milwaukee, Wis. 1980.

Stratton, B. Making the Malcolm Baldrige Award Better, *Quality Progress,* March 1990.

Taguchi, G. *Introduction to Quality Engineering.* White Plains, N.Y.: UNIPUB, 1986.

Taguchi, G. *System of Experimental Design.* White Plains, N.Y.: UNIPUB, 1987.

Thomas, D.W., et al. *Statistical Quality Control Handbook.* Indianapolis, Ind.: AT&T, 1956.

Thorpe, J. *What Every Engineer Should Know About Product Liability.* New York: Marcel Dekker, 1979.

United States Department of Commerce, 1990 Application Guidelines for the Malcolm Baldrige National Quality Award. National Institute of Standards and Technology, Gaithersburg, MD.

Index